Analytical Chemistry: A Guided Inquiry Approach

Quantitative Analysis Collection

Developed through the ANAPOGIL Project

Project Directors

Juliette Lantz

&

Renée Cole

POGIL Project
Director: Richard Moog
Associate Director: Marcy Dubroff
Publication Liaison: Sarah Rathmell

ISBN-13: 978-1-118-89131-5
ISBN: 1-118-89131-7

Table of Contents

ANA-POGIL Project Activities: Topics and Learning Goals

ANA-POGIL Activities & Analytical Texts Mapping

ANAPOGIL Project People

Leadership

Juliette Lantz, Project PI – Drew University, Madison, NJ
Renée Cole, Project co-PI – University of Iowa, Iowa City, IA

Consortium Members

Chris Bauer – University of New Hampshire, Durham, NH
Christine Dalton – Carson-Newman College, Jefferson City, TN
Anne Falke – Worcester State University, Worcester, MA
Shirley Fischer-Drowos – Widener University, Chester, PA
Caryl Fish – St. Vincent College, Latrobe, PA
David Langhus – Moravian College, Bethlehem, PA
Ruth Riter – Agnes Scott College, Decatur, GA
Carl Salter – Moravian College, Bethlehem, PA
Mary Walczak – St. Olaf College, Northfield, MN

Affiliated Authors

Kathleen Cornely – Providence College, Providence, RI
Dan King – Drexel University, Philadelphia, PA
Paul Jackson – St. Olaf College, Northfield, MN

Additional classroom testers and project participants

Carlos Garcia – University of Texas-San Antonio, TX
Gija Geme – University of Central Missouri, Warrensburg, MO
Gabrielle Haby – University of Texas-San Antonio, TX
Elizabeth Jensen – Aquinas College, Grand Rapids, MI
Delana Nivens – Armstrong Atlantic State University, Savannah, GA
David Paul – University of Arkansas-Fayetteville, AR
Josh Smith – Armstrong Atlantic State University, Savannah, GA

Additional Chemistry Education Research personnel on the project

Meagan O'Brien – University of Central Missouri, Warrensburg, MO (undergraduate researcher)
Jennifer Schmidt – University of Iowa, Iowa City, IA (graduate student)
Wendy Schatzberg – University of Iowa, Iowa City, IA (post-doctoral fellow)
Stephanie Ryan – Chicago, IL (consultant)

Advisory Board

Diane Bunce – Catholic University, Washington, D.C.
Jonathon Crowther – Ortho-Clinical Diagnostics, Inc., (Johnson & Johnson), NJ
Jeanne Pemberton – University of Arizona, Tucson, AZ

This material is based upon work supported by the National Science Foundation under Grant No. 0717492. Any opinions, findings, and conclusions or recommendations expressed in this material are those of the author(s) and do not necessarily reflect the views of the National Science Foundation.

Overview of project

The intent of the ANAPOGIL project is to provide a set of POGIL (Process Oriented Guided Inquiry Learning) materials that impart widely accepted analytical chemistry principles while engaging students as active learners, and facilitating their development in the process skills that are essential to analytical chemists in particular and valuable to scientists in general. These classroom activities can be used collectively in traditional analytical chemistry courses or used individually in a variety of other courses that include analytical chemistry concepts. To develop the initial materials, a team of ten analytical chemists from various institutions convened over a six-year period to write, test and assess these guided-inquiry materials with the full participation of two chemical education researchers as part of an NSF funded project (DUE # 0717492). During the latter years of the project, additional classroom testing was conducted at five additional institutions.

Acknowledgments

This collection of activities and supplementary materials is the result of the interactions and support of a large number of people and organizations.

- Thanks to the National Science Foundation (Grant # DUE-0717492) for its support of the project that has allowed us to accomplish so much more than we would have believed possible.

- Special thanks to the POGIL Project, and specifically Rick Moog, for getting us started on this path. This project was born out of interactions among consortium members at a POGIL National Meeting. The POGIL Project also provided the seed money for our first consortium meeting that helped us to write a successful grant proposal.

- Many thanks to the POGIL project staff who have helped us in transforming our activities into a professionally edited and presentable form that could be published.

- Thanks to our many colleagues in the POGIL community who have provided us with the opportunity to discuss our ideas with interested, stimulating, and dedicated colleagues. Many POGIL members provided guidance, insight, and support throughout the project.

- Thanks to the faculty who have used our activities in their classrooms, and special thanks to those who have provided us with feedback to continue to improve the quality of our activities.

- A great debt of thanks is due our students. They have provided their insights into the structure of the activities and helped us improve these materials.

- Drew University further supported the project through contributions of release time and computer equipment; these contributions are gratefully acknowledged.

- Finally, a special acknowledgment from JL and RC to the fabulous consortium members whose dedication and enthusiasm for the ANAPOGIL project never wavered over the several years of continued consortium tasks. This is was an incredibly hard working, generous, creative, organized, and downright fun group of analytical chemists; their collaborative efforts were key to the success of this project.

Overview of materials

The ANAPOGIL consortium has written activities that fall into six main categories generally covered in an undergraduate analytical chemistry curriculum: Analytical Tools, Statistics, Equilibrium, Chromatography and Separations, Electrochemistry, and Spectrometry. All published activities have been reviewed and classroom tested by multiple instructors at a wide variety of institutions. Each written activity includes content and process goals as well as a list of prerequisites. This information serves to help instructors compile the activities into a full curriculum. The topical treatment of concepts, as well as advanced coverage of analytical principles, can be found in the application section of each activity. Instructors can tailor coverage for a particular course by assigning different combinations of application questions.

All materials follow the constructivist learning cycle paradigm and employ concept invention where possible. All written materials are modular in nature, and they include cues for team collaboration and self-assessment. In the *exploration phase* of each activity, a team of students is presented with a model, which may be a figure, a data table, a graph, an experimental procedure, a problem-solving methodology, a schematic of a device, or any combination of these. Directed questions encourage students to deeply explore the model, especially the fundamental relationships or concepts embedded in it. *Concept development* is brought about by convergent questions, which guide students to make connections and form hypotheses, ultimately leading them toward the understanding of the desired concept. These questions model the questions scientists ask in attempting to understand new information. Term introduction typically occurs at this phase, to help students identify their newly developed concept. Students are often asked to clearly articulate the concept using complete sentences. In addition to developing communication skills, clear expression of the concept is essential for true understanding of the concept. In the *concept application* phase of the activity (which completes the learning cycle), students are asked to apply the concept to new situations. Skill exercises are one or two step exercises that provide repetition of the key questions in an identical or similar context as the model; they help to build confidence and strengthen understanding. Finally, students are asked to apply their new knowledge to solving more advanced problems. These problems require higher level thinking skills such as synthesis or analysis, may involve a multi-step approach integrating several concepts or skills, may require assumptions, and are generally more real world in nature.

In short, these activities will take students far down the path to becoming analytical chemists. Going beyond content mastery, students should gain a strong sense of what an analytical chemist does, and more importantly get to experience many aspects of how an analytical chemist does it. These activities directly scaffold every laboratory skill and technique an analytical chemist should master; moreover, the critical thinking skills and information processing can greatly empower students to choose effective data analysis routes and make sound, defensible experimental decisions.

Overview of classroom usage

These activities are mainly designed to be used in the classroom and have been implemented in settings with class periods ranging from 50 to 75 minutes. The intent is for students to work through the activities in structured Learning Teams of 3-4 students with the instructor actively facilitating student engagement and learning during class. The activities have been listed in the Table of Contents in an order that many instructors have found useful, but each activity lists the assumed prerequisite knowledge to allow the most flexibility on the part of the instructor for sequencing the use of the activities. The materials do not replace a textbook, rather they are designed to enhance the learning experience in the classroom component of the course. However, due to the learning cycle/guided inquiry design of the activities, it is recommended that students complete the activity and subsequently read the corresponding sections of the textbook. Instructors may also assign problems from the text as part of homework assignments.

Each instructor naturally implements activities in ways that are unique to that individual and institution, but we have found some common approaches that lead to more successful outcomes. We have found an implementation that includes mini-lectures and whole-class discussions in addition to the small group work to be very effective. Students may initially move slowly in this learning paradigm, but become more adept as activity usage builds and they gain insights into learning strategies and confidence in themselves and teammates. For this reason, it is important to use activities on a regular basis (daily, weekly, biweekly, etc.) so that students become accustomed to this learning environment and their role in it. We have created an instructor's guide with detailed tips on implementing the activities (including pacing guides.) Suggested implementation strategies include assigning initial directed questions in the activity as pre-class work to be discussed at the beginning of class, using the application questions as homework, providing mini-lectures to introduce background material or to address common sticking points, using whole-class discussions to compare student responses and elaborate on ideas, and providing a summary at the end of the class to highlight the learning objectives addressed. We also highly recommend that potential users attend a POGIL workshop to gain a deeper understanding of the philosophy and facilitation of guided inquiry activities.

It is important that students get some feedback on content mastery gained through use of the activities– either through classroom summaries, graded questions, or facilitator review of student work as it unfolds in the classroom. Instructors do not typically grade the activities themselves, but often focus on particular application problems and perhaps a small subset of key questions, particularly those that are summative in nature. These can be graded individually or for the group, using a Learning Team report. It is also crucial to explicitly address process skill mastery–which entails some form of evaluating student team efforts, communication skills, information processing, self-assessment (metacognition), and critical thinking as well as the more traditional problem solving strategies. The instructor's guide includes examples of several classroom strategies to assess process skills, including assessment reports and exam questions that address process skills as well as content mastery.

A two-part instructor's guide is available for these activities. Part A focuses on answer keys and instructor tips that are specific for each activity, including common sticking points and suggestions for natural break points to fit timing needs. Part B is a more general guide, with a larger focus on overall implementation and assessment strategies, including classroom procedures and materials that have been used successfully. It includes chapters on: (1) Planning for implementation and using Learning Teams, including managing expectations and structuring the course; (2) General implementation strategies, particularly for implementation in an upper level course; (3) Facilitation strategies and materials such as buy-in activities, recorder's sheets, and strategist reports; (4) Course assessment strategies, particularly for process skills; and (5) Perspectives from experienced users including sample syllabi, course descriptions, and frequently asked questions and frequently felt feelings experienced by instructors during their first implementation (and beyond).

Accuracy, Precision and Tolerance: Sorting out Glassware

Learning Objectives

Students should be able to:

Content

- Differentiate between accuracy and precision.

- Compare the tolerance of various pieces of glassware.

- Classify data in terms of the accuracy and precision.

Process

- Interpret tables of volumetric glassware tolerances. (Information Processing)

- Infer based on tabulated data. (Critical Thinking)

- Collaborate with group members. (Teamwork)

Prior knowledge

- Types of glassware for preparing solutions, including graduated cylinders, pipets, burets, and volumetric flasks.

- Definitions of accuracy and precision.

- Correct method for reading and dispensing from a buret and pipet.

- Calculate average and standard deviation for a set of data.

Further Reading

- Harris, D.C. 2010. *Quantitative Chemical Analysis,* 8th Edition, W.H. Freeman: USA, Sections 2-5 and 2-6, pp. 37-39.

Authors

Mary Walczak, Christine Dalton, and Juliette Lantz

Consider this...

Accuracy: The relationship between a measured value and the actual value. An accurate measurement will give the "true" value.

Precision: The reproducibility of a measurement. A precise measurement yields close to the same value upon repeated measurements.

Table 1 *Volume (mL) contained in 100-mL Class A Volumetric Flask*

Trial	Jamie	Kerry	Lorne
1	99.94	99.95	99.87
2	99.90	100.09	99.86
3	99.88	100.05	99.88
average (mL)	99.91	100.03	99.87
standard deviation (mL)	0.03	0.07	0.01

precision

precision

Key Questions

1. Table 1 lists three measurements of the volume of water contained in a single 100-mL volumetric flask as measured by three students. These measurements were made by weighing the mass of water contained in the volumetric flask and correcting for buoyancy. List the three students in order of increasing accuracy assuming that the true volume of the flask is known to be 100.00 mL.

2. Explain how your group determined the order in Q1.

precision.
A precise measurement yields close to the same value repeated measurements.

3. Refer back to the volumetric flask data in Table 1. List the three students in order of increasing precision.

Kerry → Jamie → Lorne

4. Explain how your group determined the order in Q3.

Consider this...

Table 2 *Volume (mL) delivered by 25-mL Class A Volumetric Pipet*

Trial	Jamie	Kerry	Lorne	Ming
1	24.94	24.83	25.15	25.10
2	24.88	24.97	24.88	25.02
3	24.81	24.88	25.07	25.12
average (mL)	24.88	24.89	25.03	25.08
standard deviation (mL)	0.07	0.07	0.14	0.05

Key Questions

5. Table 2 lists data for four students using a 25-mL volumetric pipet. Each student used the pipet to deliver 25 mL of water three times. The reported values are corrected for buoyancy. Examine each student's accuracy and precision. List the four students in order of increasing accuracy assuming that the true volume of the flask is known to be 25.00 mL. List the four students in order of increasing precision.

Student	Accuracy	Precision
Jamie	24.88	
Kerry		
Lorne		
Ming		

6. Compare Kerry's data in Tables 1 and 2. With which glassware does Kerry have better precision? How do you decide?

7. Compare the precision of student data for volumetric flasks (Table 1) with that for pipets (Table 2). Which kind of volumetric glassware has better precision? List reasons why this may be the case.

8. Is it possible to have high accuracy and low precision? If so, give an example from Table 1 or Table 2. Is it more desirable to have high accuracy or high precision? Explain.

Consider this...

Tolerance is the permissible deviation from a specified value of a structural dimension. All glassware has some tolerance for accuracy. That is, all glassware contains or delivers volumes that can be slightly different from the stated volume that is printed on the glassware. This tolerance in manufacturing is well-documented for each type and volume of glassware.

Volumetric flasks are calibrated by manufacturers "to contain" a specific volume. Volumetric flasks are marked with "TC" indicating "to contain" that volume at a stated temperature. Volumetric flasks also have manufacturing tolerances, the permissible deviation from a specified value. Table 3, below, specifies the tolerance for KIMAX Class A volumetric flasks.

Table 3 KIMAX Brand Volumetric Flasks with Standard Taper Glass Stopper, Class A, Serialized and Certified		
Capacity, mL	**Tolerance, ±mL**	**% Relative Tolerance**
10	0.02	0.20
25	0.03	0.12
50	0.05	0.10
100	0.08	0.08
200	0.10	0.05
250	0.12	0.05
500	0.20	0.04
1000	0.30	0.03
2000	0.50	0.03

Key Questions

9. Consider the 50-, 100- and 200-mL volumetric flasks in Table 3. What are the maximum and minimum volumes that could be contained in each of these flasks?

10. One way to represent the uncertainty in the volume contained in the flask is to use the % relative tolerance. What is the formula for calculating % relative tolerance for the volumetric flasks?

11. Assign the missing entries from Table 3 to the group members and calculate the % relative tolerance of each flask.

12. As a group, compare the % relative tolerance for different size volumetric flasks. What happens to the % relative tolerance as the volume of the flask increases?

13. For optimal accuracy, do you recommend choosing a larger or smaller sized flask? On what do you base this decision?

14. As a group brainstorm a list of experimental circumstances under which it would be advantageous to choose a smaller volumetric flask over a larger flask.

Consider this...

Pipets are calibrated by manufacturers "to deliver" a specific volume. Volumetric pipets are marked with "TD" indicating "to deliver" that volume at a stated temperature. Table 4, below, specifies the tolerance for KIMAX Class A pipets.

Table 4 KIMAX Brand Reusable Volumetric Pipets, Class A		
Capacity, mL	Tolerance, ±mL	% Relative Tolerance
0.5	0.006	
1	0.006	0.6
2	0.01	0.5
3	0.01	0.3
4	0.01	
5	0.02	0.4
10	0.03	0.3
15	0.03	
20	0.03	0.2
25	0.05	
50	0.08	0.2

15. *True or False:* The acceptable range of volume delivered by a KIMAX, Class A, 10-mL volumetric pipet is 9.98 to 10.02 mL. Explain your reasoning.

16. Assign the four missing % relative tolerance values in Table 4 to group members. Calculate the % relative tolerance and add the information to the Table.

17. What happens to the % relative tolerance as the volume delivered by the pipet increases?

18. *True or False:* "Volumetric glassware with the lowest tolerance will yield the most accurate measurement." Discuss this statement as a group and come to consensus. Explain your rationale in a complete and grammatically correct sentence.

19. State in your own words the definitions of accuracy, precision and tolerance. Share your definitions with your group and discuss the variations. Agree upon a single set of definitions for these three terms.

20. What is the difference between precision and accuracy? Write a grammatically correct sentence reflecting the group's consensus.

21. What do you think were the learning objectives for this activity?

22. What teamwork skill(s) did your group practice during this activity?

Applications

23. (a) Which pipet listed in Table 2 has the best precision?

(b) Which pipet in Table 4 has the lowest tolerance?

(c) What is the difference between precision and tolerance?

24. Four students determined the concentration of a stock NaOH solution by titration of potassium hydrogen phthalate (KHP). The known concentration of this solution was 0.2023 M. The titration data is presented in Table 5. List the four students in order of increasing accuracy. List the four students in order of increasing precision.

Table 5 *Sodium Hydroxide Concentration (M) as determined by titration*			
Jamie	**Kerry**	**Lorne**	**Ming**
0.1962	0.1989	0.2010	0.0984
0.1960	0.1996	0.2013	0.0981
0.1969	0.1994	0.2013	0.0984
0.1960	0.1990	0.2016	0.0996
0.1960	0.2080	0.2018	0.1059
0.1960	0.2093	0.2018	0.1157
0.1969		0.2023	0.1121
0.1954		0.2033	0.1144
0.1939		0.2037	0.0975
0.1937			0.0956
0.1927			0.1038
			0.1009
			0.1003

25. (a) Jamie needs to pipet 1 mL of bovine serum albumin solution into a test tube. Two options are available: a 1-mL volumetric pipet and a 1000-µL micropipet. Refer to Table 4 for information about volumetric pipet tolerances. Data for the micropipet appear below. Which pipet will deliver the more accurate volume? Explain your reasoning.

Micropipet Accuracy and Precision		
Volume (µL)	Accuracy ±µL	Precision ±µL
1000	8	1.5

(b) Kerry is calibrating a 1000-µL (1 mL) micropipet. The data appear below. Evaluate the data and make a statement about the accuracy of Kerry's use of the micropipet, assuming it is calibrated to deliver 1.000 x 10^3 µL. Does Kerry's use of the micropipet fall within the manufacturer's specifications?

Table 6 Kerry's Micropipet Calibration Data			
Volume (mL)			
0.9953	0.9995	0.9772	0.9627
0.9951	1.0167	0.9965	1.0049
0.9951	1.0196	1.0015	1.0038
0.9951	1.0200	1.0041	0.9977
0.9911	1.0047	0.9945	1.0079
0.9925	1.0095	0.9677	0.9998

26. Ming has to decide between two micropipets to deliver 500 µL of lysozyme solution. One is a fixed 500-µL miccropipet, and the other is a 100-1000-µL adjustable micropipet. The tolerances for both micropipets appear in the table below. Determine the % relative tolerance for each micropipet. Based on this information, evaluate the utility of using each micropipet to deliver lysozyme solution.

Table 7 Tolerances for Fixed and Adjustable Micropipets		
	Volume (µL)	**Tolerance (µL)**
Fixed Pipet	500	2
Adjustable Pipet	100	8
	1000	1.5

27. Jamie and Ming determined the percentage of acetic acid in vinegar by titration with NaOH to a phenolphthalein end point. The results of all their titrations are presented in the following histogram. The graph shows the percent acetic acid in vinegar samples for all the titrations performed by Jamie and Ming. The average percentage of acetic acid as determined by Jamie was 4.421% and by Ming was 5.009%.

a. The actual acetic acid percentage was 5.007%. Who had the better accuracy?

b. How many titrations did each student perform?

c. Who had the better precision?

d. Explain how a student's data can have high accuracy and low precision.

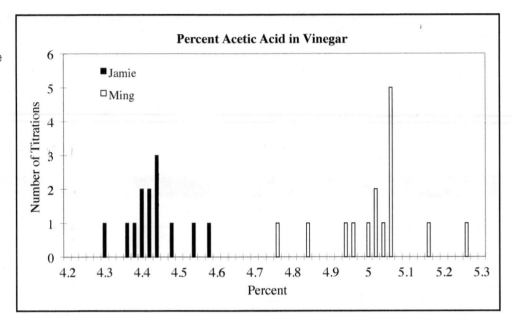

28. Frequently, analytical chemists determine concentrations in truly unknown samples. Discuss what statements about accuracy can be made in these situations.

29. Jamie is following a procedure that requires 25.00 mL of a $Pb(NO_3)_2$ solution. He needs to put the lead solution into an Erlynmeyer flask. What kind of glassware should Jamie use to measure the $Pb(NO_3)_2$ solution? Why do you make this recommendation?

30. Ming is going to pipet 10.00 mL of a sodium chloride solution. Accuracy and precision are critical to her work. She has both a 10-mL volumetric (transfer) pipet and a 10-mL measuring (Mohr) pipet. A Mohr pipet has graduation marks on it and can be used to transfer various volumes.

a. The tolerances for Ming's 10-mL Mohr pipet are given in Table 8. Tolerances for volumetric pipets are in Table 4. Which pipet has the lower tolerance? Which pipet should Ming use?

Table 8 KIMAX Brand Reusable Measuring (Mohr) Pipets, Class A, Color Coded		
Capacity, mL	**Subdivisions, mL**	**Tolerance, ±mL**
1	0.1	0.010
1	0.01	0.010
2	0.1	0.01
5	0.1	0.02
10	0.1	0.03
25	0.1	0.05

b. Ming wanted to make sure she used the more precise pipet. Therefore, she used each pipet to deliver 10.00 mL of water, weighed each delivery, and converted to volume. Her results are shown in Table 9. Based on the data provided, which pipet should Ming use for her laboratory work?

Table 9 Ming's Pipetting Data		
	Volumetric Pipet	**Measuring Pipet**
Trial	**Volume (mL)**	**Volume (mL)**
1	10.01	9.98
2	10.02	9.96
3	10.02	10.03
4	10.02	10.02
5	9.99	9.97
6	10.00	9.95
7	9.99	9.96
8	9.99	10.03

c. If your recommendations in parts (a) and (b) above are different suggest reasons why this might be the case.

31. Data for calibrating a 100-mL and a 25-mL volumetric flask are shown in Table 10. Which flask truly has the best precision?

Table 10	Volumes obtained when calibrating 100-mL and 25-mL volumetric flasks.		
100 mL flask **Volume (mL)**		**25 mL flask** **Volume (mL)**	
99.54	**100.03**	**24.64**	**25.03**
100.04	100.00	25.04	25.01
99.99	99.99	25.00	24.99
100.04	100.00	25.04	25.01
99.96	100.07	24.97	25.06
100.02	99.97	25.02	24.98
100.02	99.99	25.02	25.00

32. Lorne and Kerry are working on preparing Copper Atomic Absorption standards in the 2-8 ppm range. They are working with a 100.00±0.01 ppm primary standard solution. Kerry decides to use volumetric pipets and flasks to make 100 mL of a 5 ppm Cu secondary standard. Lorne decides to make the solution gravimetrically using a top loader balance (±0.01 g). Assuming that Kerry and Lorne use the equipment properly, which method results in a solution with better precision?

Solutions and Dilutions

Learning Objectives

Students should be able to:

Content

- Design a procedure for making a particular solution and assess the advantages of different approaches.

- Choose the appropriate glassware to ensure the desired level of precision of a particular solution.

- Convert between different concentration units (e.g., ppm to M).

Process

- Develop alternative pathways for diluting solutions. (Information processing)

- Design approaches for preparing solutions. (Problem solving)

- Infer chemical processes based on reactions. (Critical thinking)

Prior knowledge

- Types of glassware for preparing solutions, including graduated cylinders, pipettes, burets, and volumetric flasks.

- Correct method for reading and dispensing from a graduated cylinder, pipette, buret, and volumetric flask.

- Definitions of primary standard, secondary standard, calibration standard, accuracy and precision, ppm, molarity.

Further Reading

- Harris, D.C. 2010. Quantitative Chemical Analysis, 8th Edition, New York: W.H. Freeman. pp. 37-39.

- Skoog, D.A., D.M. West, F.J. Holler, and S.R. 2004. Crouch, Fundamentals of Analytical Chemistry, 8th Edition, Thompson Brooks/Cole: USA, Sections 13B and 13C, pp. 340-343.

- Louisiana Universities Marine Consortium (LUMCON) Bayouside Classroom, dissolved oxygen section, available at: http://www.lumcon.edu/education/studentdatabase/dissolvedoxygen.asp

Authors

Mary Walczak and Christine Dalton

PGLANA004002a

POGIL

Consider this...

Shelby needs to make a 3.00 M acetic acid solution from an acetic acid standardized solution that is 15.0 M. The analytical equipment available to Shelby in the lab includes the volumetric flasks and pipets shown below. Although several different size volumetric flasks are available, Shelby opts to use the 100-mL flask because it is both clean and dry.

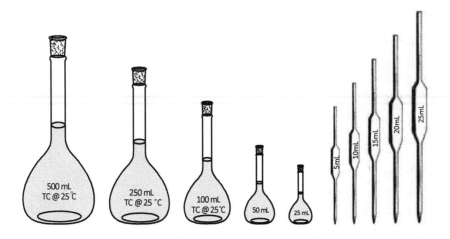

Key Questions

1. What size pipet does Shelby need to use to make 100 mL of 3.00 M acetic acid solution? Devise a general mathematical expression for calculating the concentration of the resulting solution.

   $$C_1V_1 = C_2V_2$$
   $$(15M)V_1 = (3M)(100mL)$$
   $$V_1 = 20 \ mL$$
   $$\downarrow$$
   $$20 \ pipet$$

2. The dilution factor (initial volume of solution/final volume of solution) is a way of expressing the extent to which a solution is diluted. What dilution factor is used to prepare the solution described in Q1?

   $$\frac{20mL}{100mL} = \frac{1}{5}$$

3. Shelby is not concerned about the dryness of the pipets. Explain, based on proper pipetting techniques and your laboratory experiences, exactly how Shelby should use the pipet. Why does proper lab technique eliminate the need for dry pipets?

   Shelby could rinse the pipette with tiny amount of the concentration solution 3 times to ensure that droplets of that solution are inside of pipette before we start to measuring out solution to trasfer

4. Was it necessary for the volumetric flask to be dry? Discuss as a group and explain your answer and reasoning in a grammatically complete sentence.

Yes,

5. Assign different volumetric flasks to group members. For each size flask, determine the volume necessary to prepare that volume of a 3.00 M acetic acid solution from the 15.0 M standard solution. Determine the dilution factor used to prepare any of these solutions. Check your group's answers by comparing with other groups. Resolve any differences.

6. Suppose Shelby's supervisor said to make 250 mL of 3.00 M acetic acid solution. How would you suggest Shelby prepare this solution?

$$C_1 V_1 = C_2 V_2$$
$$(15 M)V_1 = (3 M)(250 mL)$$
$$V_1 = 50 mL$$

7. Shelby's supervisor asked for 100 mL of a 2.00 M acetic acid solution. Could Shelby prepare this from the 15.0 M standard solution using the available glassware in Model 1? If so, how? If not, what other glassware might Shelby want to use?

$$(15 M)V_1 = (2 M)(100 mL)$$
$$V_1 = 13.33 mL$$

Consider this…

Making Solutions: *Rules of Thumb*

- Graduated cylinders are considerably less accurate and precise than glass transfer pipets.

- Dilution in one step is better than two.

- Larger glassware has less relative uncertainty.

- Measuring (Mohr) pipets are less precise than glass transfer pipettes.

- Waste handling is expensive.

- Glassware is designed to hold a specific volume only at a stated temperature.

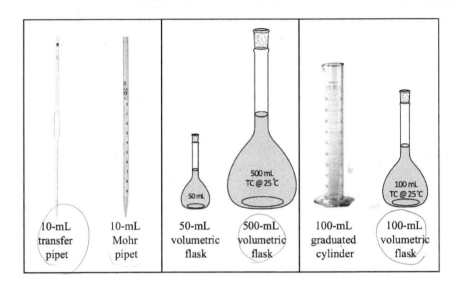

Key Questions

8. Considering the "rules of thumb" listed above, circle the glassware in each pair above that will provide the lower uncertainty.

9. Under what conditions is it advantageous to use the smaller volumetric flask in the center panel above? When is the larger flask the optimum choice? Compare answers with group members and arrive at a consensus.

Use the smaller flask, when the solution or agent is more expensive, so it is less cost to make less solution. The large flask is for large amount like water, or for precision is important, beause it has a lower relative uncertainty *handwritten top:* $C(5) = 5(1000)$

Consider this...

Reagan is doing an atomic absorption experiment that requires a set of zinc standards in the 0.4-1.6 ppm range. A 1000 ppm Zn solution was prepared by dissolving the necessary amount of solid $Zn(NO_3)_2$ in water. The standards can be prepared by diluting the 1000 ppm Zn solution. Table 1 shows one possible set of serial dilutions (stepwise dilution of a solution) that Reagan could perform to make the necessary standards. Solution A was prepared by diluting 5.00 mL of the 1000 ppm Zn standard to 50.00 mL. Solutions C-E are called "calibration standards" because they will be used to calibrate the atomic absorption spectrometer.

handwritten:
$5(1000) = 50 C_2$
$5000 = 50 C_2$
$C_2 = 100$

$C_1(5) = C_2(100)$
$(1000)5 =$
$5000 = C_2(100)$

Table 1 *Dilutions of Zinc Solutions*

Solution	Zinc Solution Concentration (ppm Zn)	Volume used (mL)	Diluted Volume (mL)	Solution Concentration (ppm Zn)	Solution Concentration (ppm $Zn(NO_3)_2$)	Solution Concentration (M $Zn(NO_3)_2$)	Solution Concentration (M Zn)
A	1000	5.00	50.00	1.00×10^2	2.90×10^2	1.53×10^{-3}	1.53×10^{-3}
B	Solution A	5.00	100.00	5.00	14.0		
C	Solution B	5.00	50.00	0.50	1.00	7.65×10^{-6}	
D	Solution B	10.00	50.00	1.000	3.00		
E	Solution B	25.00	100.00	0.25	4.000		1.911×10^{-5}

handwritten table annotations: "Calibration standards" bracket next to C, D, E; "10 0" under ppm Zn for A; C_1(100), V_1, V_2 labels on row B; (5.00) under Solution B on C; 0.50 under Solution B on D

Key Questions

handwritten right side:
$(5.00 \text{ ppm})(5.00 \text{ mL}) = C_2(50 \text{ mL})$
$25 \text{ ppm} = C_2(50)$
$C_2 = 0.50 \text{ (Solutn C)}$

$(1)25 = C_2 100$
$C_2 = 0.25$

10. Using your general scheme for calculating the concentration of diluted solutions devised in Q1, calculate the resulting concentration (ppm Zn) for each solution A-E above. Enter your answers into the diagram in Figure 1 and into the boxes in Table 1. Verify answers provided in Table 1.

handwritten:
$C_1 V_1 = C_2 V_2$
$(1000 \text{ ppm})(5 \text{ mL}) = C_2(50 \text{ mL})$
$5000 \text{ ppm} = C_2(50)$
$C_2 = 100 \rightarrow A$
$100 \text{ ppm}(5 \text{ mL}) = C_2(100)$
$500 = C_2 100$
$C_2 = 5.00 \rightarrow B$

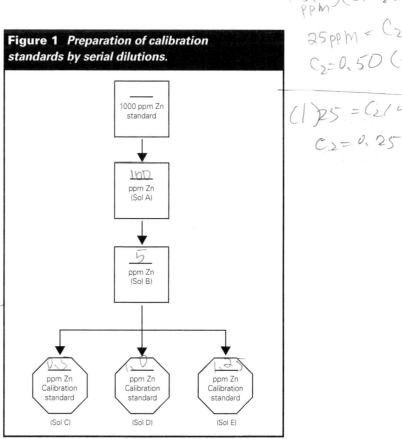

Figure 1 *Preparation of calibration standards by serial dilutions.*

1000 ppm Zn standard

↓

100 ppm Zn (Sol A)

↓

5 ppm Zn (Sol B)

↓

0.5 ppm Zn Calibration standard (Sol C) | 1.0 ppm Zn Calibration standard (Sol D) | 1.25 ppm Zn Calibration standard (Sol E)

11. The atomic mass of Zn is 65.4094 amu and the molar mass of $Zn(NO_3)_2$ is 189.4194 amu. Devise a scheme to calculate the solution concentrations in units of ppm $Zn(NO_3)_2$. Calculate the resulting solution concentrations in ppm $Zn(NO_3)_2$ for Solutions A-E. Enter your results in Table 1.

$$1000 \text{ ppm Zn} \left| \frac{1 \text{ mol Zn}}{65.409 \text{ g Zn}} \right| \frac{1 \text{ mol Zn(NO}_3)_2}{1 \text{ mol Zn}} \left| \frac{189.4194 \text{ g Zn(NO}_3)_2}{1 \text{ mol Zn(NO}_3)_2} \right.$$

$$= 2896 \text{ ppm Zn(NO}_3)_2$$

12. Devise a scheme to calculate the solution concentrations in units of molarity (M $Zn(NO_3)_2$) and (M Zn). Calculate the resulting solution concentrations in M $Zn(NO_3)_2$ and M Zn for Solutions A-E. Enter your results in Table 1.

13. Compare the solution concentrations expressed as ppm Zn and ppm $Zn(NO_3)_2$. Compare the concentrations expressed as M Zn and M $Zn(NO_3)_2$.

Which units allow easy conversion between chemical species (e.g., Zn and $Zn(NO_3)_2$)?

Which units express concentrations in numbers with easily expressed magnitudes?

Suppose you have an analyte for which you don't know the molar mass. Which concentration units would you use?

14. For the concentrations involved in this particular experiment, which set of units is most convenient? Explain your reasoning.

15. Write a laboratory procedure for the preparation of Solution B from Solution A for Reagan to carry out.

16. Although only 50 mL of each calibration standard (Solutions C-E) are needed for the experiment, Reagan made 100 mL of Solution E. Give a rationale for this decision.

17. Suppose Reagan made an error in preparing Solution A and ended up with a 100.5 ppm Zn solution. What is the impact of this error on Solutions B-E?

Consider this...

Alternatively, Reagan could prepare the same calibration standards by first preparing a 50 ppm standard and then diluting this standard to make the three calibration standards, as shown in Figure 2. The advantage to this approach is that only one dilution of the 1000 ppm Zn standard is needed before preparing the calibration standards.

18. Brainstorm ideas about how the 50 ppm standard can be prepared. As a group, come to a consensus on a dilution scheme to prepare this standard and explain the rationale behind choosing this particular method.

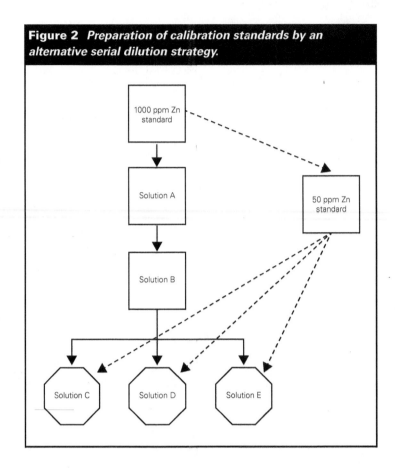

Figure 2 *Preparation of calibration standards by an alternative serial dilution strategy.*

19. Divide the calibration standards, Solutions C, D, and E, among group members. For each calibration standard, determine how the 50 ppm standard could be diluted using the volumetric flasks and pipets available (see Q1) to prepare the solution. The calibration standard concentrations are the same as in Figure 1 and 2 Solutions C, D and E.

20. Compare the dilution schemes of Figures 1 and 2. Which is the better method in terms of the following parameters?

a. Time

b. Error

c. Waste generation

d. Amount of glassware requiring cleaning

Based on your answers, which model is better and why?

21. Reagan decides another calibration standard is needed between 0.5 and 1.0 ppm. Using Figure 1 or 2 (whichever you decided was better in Q20), how would you make a 0.75 ppm Zn calibration standard (Solution F)?

22. List important considerations when developing a procedure to make a solution.

23. Explain the process of designing an experiment in which calibration standards (solutions used to calibrate an instrument) must be made from a standard solution.

Applications

24. List the advantages and disadvantages of preparing a 1:1000 dilution in a single step. Consider time, resources and error.

25. A bottle is labeled 56.2 ppm $FeCl_3$. Express this concentration in ppm Fe^{3+} and in molarity.

26. If the lab equipment in Q1 is available for use, explain how you would prepare a 0.02000 M potassium permanganate ($KMnO_4$) solution from a standardized $KMnO_4$ solution with a concentration of 0.1000 M. Write a lab procedure for preparing the 0.02000 M solution.

27. Potassium dichromate, $K_2Cr_2O_7$ is a carcinogen, however it still plays a useful chemistry role in some procedures. Suppose you need to make a 0.1000 M $K_2Cr_2O_7$ solution from a 1.000 M stock solution. Using glassware in Q1, how would you prepare the smallest quantity of this 0.2000 M solution?

28. Shelby and Reagan have been hired by a forensic science lab to analyze trace metal content in crime scene evidence. Their current project involves determining zinc content in human hair. Shelby and Reagan prepare a nitric acid solution to dissolve the hair using a process called digestion. The nitric acid oxidizes the keratin protein. Shelby and Reagan need 100 mL of an aqueous 0.3 M HNO_3 solution for this procedure. They remember their chemistry training and plan to "add acid to water," and add the necessary volume of concentrated (16.0 M) HNO_3 to a sufficient quantity of water.

Reagan decides to use a 100-mL volumetric flask to prepare the solution. The flask is filled about three-quarters full of deionized water and the acid is measured using a 10-mL graduated cylinder. After mixing the solution well, Reagan fills the flask with enough deionized water to bring the level to the mark on the neck and mixes the solution well again. Shelby, on the other hand, uses a 100-mL graduated cylinder to measure 100 mL of deionized water, pours it into a beaker, adds the HNO_3 from a 10-mL graduated cylinder and stirs well.

a. How many milliliters of the concentrated HNO_3 are needed to make 100 mL of 0.3 M solution?

b. What advantages are there to using a volumetric flask to prepare the HNO_3 solution? Explain.

c. What advantages are there to using the graduated cylinder to prepare the HNO_3 solution? Explain.

d. Consider the photograph of a 10-mL graduated cylinder at the right. Estimate the uncertainty in the volume of concentrated HNO_3 that Shelby and Reagan measure using this glassware.

http://homepages.ius.edu/DSPURLOC/c121/week3A.htm

e. As they are working in the lab, Shelby and Reagan notice the different approaches they have taken to making this solution. What are the advantages and disadvantages of each approach? Which student made the better choice? Why?

f. Considering the role of the nitric acid solution in the hair analysis, how important is the accuracy of the concentration?

g. If another student asked you for advice on what glassware to use in preparing the HNO_3 solution, what would you tell him or her?

29. In order to measure the amount of Zn in hair using an atomic absorption instrument Shelby needs to prepare a calibration curve of atomic absorbance for five solutions of known Zn concentration. For each concentration 100 mL of solution is needed and the standards need to be in the 0.4 – 1.6 ppm Zn range. The available glassware is listed in Q1 and a 1000 ppm Zn primary standard solution is available.

Table 2   Capacity and Tolerance of Available Graduated Cylinders, Volumetric Flasks and Pipettes

Graduated Cylinders		Volumetric Flasks		Volumetric Pipets	
Capacity, mL	Tolerance, ±mL	Capacity, mL	Tolerance, ±mL	Capacity, mL	Tolerance, ±mL
10	0.05	50	0.05	10	0.03
100	0.5	100	0.08	15	0.03
250	1	200	0.10	20	0.03

a. Which glassware would be the best to use to prepare the calibration standards?

b. Describe how you would make five calibration standards in the 0.4–1.6 ppm range starting with the 1000 ppm Zn primary standard solution.

c. Explain, using complete sentences, why the glassware chosen for this problem is the best of those listed in Table 2.

30. What type of glassware should be used for making up solutions when you need to know concentrations accurately? Under what circumstances can graduated cylinders be used to prepare solutions?

31. State in your own words why the precision of the zinc solution in Q29 is important, but the nitric acid solution precision in Q28 is not.

32. Shelby needs to perform dilutions of a 52.08 ppm $CaCl_2$ solution to prepare a set of calibration standards for atomic absorption spectroscopy. The Ca working range for AA is 1-4 ppm Ca. The pipettes and volumetric flasks in Q1 are available for use.

    a. Convert the concentration of the 52.08 ppm $CaCl_2$ solution to units of ppm Ca and in molarity.

    b. Explore combinations of available pipets and volumetric flasks to determine how to dilute the 52.08 ppm $CaCl_2$ solution to make five calibration standards that fall into the 1-4 ppm Ca range. Make a table showing how to prepare the five calibration standards and the resulting concentrations (ppm Ca).

    c. Write a set of instructions for a technician to follow when preparing these calibration standards. Be specific about which glassware should be used.

    d. Under what circumstances would you opt for the next smaller size volumetric flask than you chose in question Q32b and Q32c? When would you choose the next larger size volumetric flask?

33. Shelby is going to analyze drinking water samples for sodium and calcium using inductively coupled plasma – atomic emission spectroscopy. Calibration standards containing both NaCl and $CaCO_3$ within the 1–10 ppm range are needed. Shelby wants to prepare 100 mL of each calibration standard in order to ensure there is enough solution. Explain how Shelby should prepare a primary standard solution containing 100 ppm of each element from primary standard grade solid NaCl and $CaCO_3$. How should Shelby prepare eight 100 mL calibration standards within the range given? Do the calculations using a spreadsheet program. In your narrative response comment on the types of glassware used to prepare the solution.

34. Shelby is measuring chloride concentration in environmental samples using a chloride ion selective electrode. In order to calibrate the electrode a set of five chloride standards in the 10^{-1} to 10^{-5} M range is needed. Shelby has a standard solution of 1.000 M NaCl to use in preparing the standards. The glassware available in the lab includes an ample supply of all the volumetric flasks and pipettes displayed in Q1.

    a. Shelby pipets 10 mL of the 1.000 M NaCl solution into the 100-mL volumetric flask, dilutes to volume and mixes the solution by inverting the flask 30 times. What is the concentration of the resulting solution?

    b. What volume of the 1.000 M NaCl solution is required to make 100 mL of the 10^{-5} M chloride solution? How do you suggest Shelby measure this volume?

    c. Calculate the volume required for the three remaining concentrations. What practical difficulties might be encountered when preparing these solutions?

    d. List ideas about how Shelby might prepare these other three solutions using the available equipment.

    e. Develop a strategy for Shelby to prepare the five chloride standards. Write instructions for a lab technician to use in preparing the standards necessary for this analysis.

35. Sodium hydroxide is a hygroscopic solid, meaning that it absorbs moisture from the atmosphere. Consequently, if the concentration of a NaOH solution must be known precisely, it must be standardized against an acid. Furthermore, the acid must be a primary standard, a substance that can be weighed precisely resulting in a well-known amount of acid. This process, called standardization, is a titration of the known quantity of acid with the NaOH solution. A commonly used primary standard weak acid for standardizing NaOH solutions is potassium acid phthalate, often abbreviated "KHP."

a. If 2.61 g NaOH was dissolved in 1 L of deionized water, estimate the resulting NaOH solution concentration.

b. What type of laboratory glassware would you use to prepare this NaOH solution? Explain your rationale.

c. Titrations of KHP with the NaOH solution were performed to determine the actual NaOH concentration. Data for titration of KHP with NaOH are provided below. Use the data to determine the actual concentration of the NaOH solution. Comment on the quality of the data provided.

Mass KHP (g)	Volume NaOH solution (mL)
0.2432	19.10
0.2358	18.66
0.2327	18.63

d. Given the concentration determined by titration and the expected concentration calculated in part a, comment on the purity of the weighed NaOH sample.

36. The wet chemical method for measuring dissolved oxygen in water was developed by Winkler in 1888 and is still in use today. This method involves oxidation of iodide ion to iodine by the oxygen in the sample. The amount of iodine produced is determined by titration with a sodium thiosulfate solution. Since the titration is often not conducted at the time of sampling, the dissolved oxygen in a collected sample is "fixed" by precipitation as manganese (III) hydroxide ($Mn(OH)_3$). This stabilizes the oxygen from the water sample until a later time when the titration in performed. Overall, the stoichiometric reactions involved are:

Reaction	Purpose
1. $Mn^{2+} + 2\,OH^- \rightarrow Mn(OH)_2$	Mn^{2+} is loosely bound to hydroxide in basic solution
2. $2Mn(OH)_2 + \frac{1}{2}\,O_2 + H_2O \rightarrow$ $2Mn(OH)_3$	Mn^{2+} is oxidized to Mn^{3+} in the presence of strong base and fixes oxygen from the water sample
3. $2Mn(OH)_3 + 2I^- + 6H^+ \rightarrow$ $2Mn^{2+} + I_2 + 6H_2O$	Iodine (I_2) is produced upon acidification of the sample with stoichiometry of one I_2 molecule for every oxygen atom
4. $I^- + I_2 \rightarrow I_3^-$	Iodine complexes with excess I^- to form triiodide (I_3^-)
5. $I_3^- + 2S_2O_3^{2-} \rightarrow 3I^- + S_4O_6^{2-}$	Triiodide is reduced to iodide with thiosulfate via titration using starch indicator

The reagents necessary for the procedure include:

3 M Manganese (II) chloride
4 M sodium iodide *and* 8 M sodium hydroxide
50% v/v sulfuric acid
1.0 g/1000 mL starch indicator solution
0.018 M sodium thiosulfate solution.

    a. For each reaction 2-5 above, identify the stoichiometric relationship between the analyte of interest (oxygen) and the species in each reaction.

    b. Considering the relationships explored in Q31 above, which of the five solutions listed above must be know to a precise concentration (e.g., 3 or 4 significant digits)? Which solutions need only approximate concentrations? Explain how you arrived at your answers.

37. Iron can be determined spectrophotometrically after reaction with a ligand that results in a highly colored complex. In one determination, iron is reduced with hydroquinone to ensure it is all in the Fe^{2+} oxidation state. Then, o-phenanthroline is added to complex with the Fe^{2+}. The reactions are:

The reagents necessary to perform this experiment include a solution of hydroquinone, citric acid buffer, a solution of o-phenanthroline, and an $Fe(NH_4)_2(SO_4)_2 \bullet 6H_2O$ solution. The concentrations of which of these solutions must be known precisely in order to make an accurate determination of iron in an unknown sample? Explain your rationale.

38. Shelby is measuring the concentration of lead (as Pb^{2+}) in paint samples using an atomic absorption instrument. A primary standard solution of aqueous $Pb(NO_3)_2$ is used to prepare standards for calibrating the instrument. Shelby weighs out 0.2500 g of primary standard grade $Pb(NO_3)_2$, quantitatively transfers it to a container and dissolves it in DI water for a total solution volume of 250 mL. Various pieces of glassware are available including a 250 ± 0.8 mL graduated cylinder and 250 ± 0.12 mL volumetric flask.

   a. Calculate the solution concentration if the 0.2500g of $Pb(NO_3)_2$ is dissolved to a total of 250 mL of solution. Determine the concentration in units of ppm Pb and ppm $Pb(NO_3)_2$.

   b. Estimate the uncertainty associated with preparing the solution using (1) the volumetric flask and (2) the graduated cylinder. Report the uncertainty as a range of concentrations (ppm Pb). Compare results.

   c. What do these ranges of concentrations reveal about the difference in accuracy between the graduated cylinder and the volumetric flask? Explain, using results from part b above, why volumetric flasks should always be chosen over graduated cylinders for analytical methods.

Classical Analytical Methods: A Design Perspective on Volumetric Measurement

Learning Objectives

Students should be able to:

Content Goals

- Describe the characteristics of the chemical substances and reactions used in volumetric analysis, particularly as they affect accuracy and precision.

- Explain the rationale for the procedural steps and techniques in volumetric analysis, particularly as they affect accuracy and precision.

- Explain how visual indicators are used.

- Explain the role of primary standard materials.

Process Goals

- Apply chemical concepts and knowledge of mass and volume measurement to making design decisions about volumetric measurement. (Critical Thinking)

- Use typical titration measurement calculations. (Problem Solving)

Prior knowledge

- Conceptual introduction to titration.

- Experience with doing a titration of any type.

- Glassware tolerance and proper usage procedures.

- Concepts of accuracy and precision.

Author

Christopher F. Bauer

PGLANA004003a POGIL

Consider this...

To the right is a typical set up for conducting a titration.

The materials include a buret containing a titrant solution of known concentration, and a receiving flask which contains a sample solution in which the analyte of interest is present at an unknown concentration. (Other variations are possible but for right now, we'll keep it simple.)

The process involves adding titrant slowly (drop-by-drop or "dropwise"). The titrant reacts with the analyte. When just enough titrant is added to use up all the analyte (called the point of equivalence), no more titrant is added. The volume of titrant added is determined by the change in the height of the liquid in the buret. Since a buret is calibrated in units of volume, the volume added is obtained.

An example of a titration reaction for titrant T and analyte An is

$$2\,T + An \rightarrow Product$$

Assume the materials include 0.2000 M titrant and 20.00 mL of 0.2000 M analyte.

Key Questions

1. For the information given, calculate the volume of titrant that would be needed to reach the equivalence point.

2. Convert your calculation into a general equation that shows how volume of titrant V_t is calculated from the concentrations of analyte [An] and titrant [T].

3. If this were a real titration analysis, which quantity in Q2 would you not know?

4. The titration described here is intended to determine the concentration of analyte [An] as the unknown in the sample. Rearrange your equation in Q2 so [An] is the final result.

5. How many significant figures are in each of the numerical values used in Q4? How many significant figures should be in the final result for [An], as determined in an actual titration?

In order for the calculated molarity of analyte to be accurate, you have to guarantee that each of the quantities in the calculation in Q4 can be determined accurately. The next four sections ask questions about the validity of each of the numbers you use for your calculation—volume of sample solution, concentration of titrant solution, volume of titrant, stoichiometric coefficients.

Consider this...

Volume of sample solution

6. List several types of lab glassware available for volume measurement.

7. Which type of glassware is the best for transferring one specific volume of the analyte solution into the titration flask? Justify what you mean by "best".

8. Brainstorm in your group. Describe at least three things (i.e. aspects of lab technique) you would do to ensure that the pipet will in fact deliver 20.00 mL accurately.

9. Convert your list into a sequence of procedural steps that should be routinely done to assure accurate delivery of a volume. Then compare your procedure with your text or lab recommendations.

Consider this…

Concentration of titrant solution

10. The concentration of titrant is stated to be 0.2000M. Assume it was prepared by dissolving a mass of solid T in an appropriate volume of water. The conditions listed in the following table must be met to guarantee that four significant figures are justified. Split the items among group members. Each person should propose a sensible reason for why that condition must be met. When everyone is ready, the group should share and discuss the reasons, accepting or improving each one.

	Conditions to be met	Reason for meeting this condition
A	Select appropriate type of flask in which to make solution.	
B	Last rinse water drains in sheets and does not leave droplets behind.	
C	Titrant that is a pure substance.	
D	Titrant that has a known molar mass.	
E	Titrant does not absorb moisture or can be dried.	
F	Quantity of solid that is massed to make solution is larger than 100 mg.	
G	Balance is tared, leveled, and protected from drafts.	
H	Careful transfer of solid into flask.	
I	Rinsing of residual solid from container used to measure mass.	
J	Rinsing of neck of receiving flask.	
K	Add some solvent to flask and totally dissolve solid.	
L	Fill to near maximum, then mix well.	
M	Fill to "the line" and mix.	

Primary Standards are chemical substances that fulfill many of the desirable characteristics discussed in the table—nearly 100% pure; stable to air, moisture, and drying; long shelf-life; ease of handling. Large molar mass is also desirable because, for a given number of moles, more mass will be taken, and taking a larger mass reduces the relative error of mass measurement. With care, mass measurement of at least 100 mg provides a highly accurate quantity. Thus solutions made from this mass can have concentrations known with high accuracy. Thus, these are ideal materials to use for making titrants. Many titration procedures require a direct or indirect link to a primary standard.

The most common primary standards include:
- sodium carbonate (Na_2CO_3) for titration of acids
- potassium hydrogen phthalate ($KHC_8H_4O_4$) for titration of bases (called KHP)
- potassium dichromate ($K_2Cr_2O_7$) for oxidation-reduction titrations

Refer to your text or information provided by your instructor for further detail.

11. The concentration of titrant [T] will have high accuracy if all of the conditions listed in the table above are met. We can speak of these conditions as falling into a few categories listed below. Put the individual conditions (A through M) into the category or categories that fit best. (Just list the letters)

Instruments perform as designed   _____

"Quantitative" transfer of material   _____

Uniform concentration of titrant solution _____

What you weigh is what you get   _____

12. Assign each group member one category. Give one specific example of how [T] could be incorrect. Do this by completing the sentence: "The concentration of titrant would be (consistently too high, consistently too low, more inconsistent) than it should be if ......" Share your sentences with the group and determine whether they make sense. Write them down.

Consider this...

Volume of titrant added

The model on an earlier page shows a picture of a buret, which is used to deliver the titrant into the flask containing the analyte. To obtain a value for the volume of titrant added, you have to know when to stop adding titrant, and you need to determine the delivered volume.

There are two typical ways to know when to stop adding titrant—look for a color change caused by a substance added to the analyte solution or follow the concentration of one of the species using a chemical probe (such as a pH electrode). Color change indicators are considered later in this activity. Chemical probes are considered in another activity.

The volume delivered is measured simply as the difference between the initial volume level and final volume level of titrant in the buret. Each of these volume measurements may be estimated using the markings on the buret to the nearest 0.01 mL.

13. Use propagation of error procedures to calculate the combined volume reading error for the volume delivered.

14. Assume you are using a 50.00-mL buret. Each person in your group should calculate the percent error that "volume difference" contributes if the total volume of titrant needed is 5.00 mL, 15.00 mL, 40.00 mL, and 80.00 mL. (The last one requires refilling the buret.)

Volume used	Percent Error
5.00 mL	
15.00 mL	
40.00 mL	
80.00 mL	

15. Based on your result in Q14, what recommendation would you make regarding a desirable volume of titrant to use in an analysis? Explain your answer.

16. Brainstorm in your group to identify at least three important aspects of procedure (i.e. lab technique) that will guarantee that the buret will in fact deliver 40.00 mL accurately.

17. Convert your list into a sequence of procedural steps that should be routinely done to assure exact delivery of a volume. Then compare your procedure with your text or lab recommendations.

The *equivalence point* is the volume of titrant one would expect to add so that the moles of titrant added balances the moles of analyte present according to the chemical reaction occurring between them. The *end point* is the volume of titrant added up to the point of change in the indicator color or in the chemical probe signal. In a good titration procedure, the end point must match the equivalence point closely.

---

18. If the endpoint happens too soon relative to the equivalence point, what will happen to the calculated concentration of analyte [An] (revisit key Q4)? If the endpoint happens too late?

Consider this…

Stoichiometric Coefficients

You have discussed the validity of three of the four quantities used in the titration calculation in key Q4 (volume V_t and concentration [T] of titrant, and volume of sample V_{sample}). The last quantity concerns reaction stoichiometry (mole ratio of analyte to titrant).

19. Give each group member one of the four questions (a,b,c,d) below. Think about the implications. Then convert this into a procedural suggestion—write a grammatically correct sentence that completes the phrase "Titration reaction must …" Share your answers with the group. Agree or improve upon the statements written by individuals. Write those down.

    a. If you do not know what the chemical reaction between titrant and analyte is, can you complete the calculation in key Q4? Why?

    b. Think about doing a titration. Assume you are near the endpoint and are adding titrant one drop at a time so as not to overshoot the endpoint. How annoying would it be if you had to wait 10 minutes with each drop to see if the endpoint has been reached?

    c. All chemical reactions are equilibrium systems and are described by an equilibrium constant. Would you want the equilibrium constant to have any particular range of values? Why?

    d. T and An are not the only chemical substances present in the titrant solution and the sample solution. When the titrant and analyte solutions and everything they contain are mixed, what must happen and what must not happen in order for your calculation in Q4 to be valid?.

20. Analytical chemists talk about titration systems as needing to have the qualities of being known, exact, complete, and rapid. Match each of these adjectives to one or more of the questions above.

Adjective	Item (a, b, c, d) that matches
Known	
Exact	
Complete	
Rapid	

21. Each group member should identify one new thing learned about the design of titration procedures. Write down those things.

22. If you were asked to design a titration method, how confident are you that you would be aware of the things you need to think about to do that well? Discuss this as a group and make a list of any specific issues that you are still unsure about.

Application

23. Return to the issue of finding the equivalence point (review question 18). Assume you are using a colorimetric indicator (one which changes color at the equivalence point). One way in which such an indicator (In) can work is that it undergoes a reaction with substance T itself, and that reaction leads to a product that has a different color than the unreacted indicator. In other words, a reaction like this occurs:

In (color A) + T→ Product (color B)

Since the Indicator is typically added to the analyzed sample (in the flask) before the titration is begun, Indicator is present all the time. This seems to be a contradiction regarding the general principle you identified in Q19d. In choosing and using indicators then, there are strategies one can employ to make sure that problems are minimized so as not to affect measurement accuracy. Make of list of your suggestions. Hint: Think about equilibrium and concentrations.

24. In typical acid/base titrations, the endpoint is often based on a visual indicator whose color changes in response to protonation or deprotonation because of the rapid change of pH at the equivalence point. This can be explained as a Le Chatelier effect on the indicator equilibrium:

H-In (Color A) $\rightleftarrows$ In (Color B) + H⁺

This means that the change from one color to the other occurs at a pH around the pK_a of the indicator substance. The equivalence point of each of the following titration systems is shown in the table below. Information is also provided on several acid/base indicators along with their pK_a values. Choose the indicator that will be the most accurate for the titration system shown, and indicate the color change that will be observed during the titration.

	pK_a	acidic form color	basic form color
Methyl red	5.0	red	yellow
Bromthymol blue	7.1	yellow	blue
Phenolphthalein	9.0	colorless	pink

Titrant solution	Analyte solution	pH range of equivalent point	Selected Indicator	Direction of color change
Strong acid	Strong base			
Strong acid	Weak base			
Strong base	Strong acid			
Strong base	Weak acid			

25. If you were designing a titration based on a reaction that no one had ever used before in a titration, what "homework" would you need to do?

26. Sometimes it is advantageous to perform what is called a "back titration". Two steps are used:

Step 1: Add an excess of reagent R to react with analyte completely

Step 2: Titrate the excess reagent R with an appropriate titrant

This obviously takes extra work. Under what circumstances might this approach be preferred over a direct titration? What additional information would you need to know in order to achieve accurate analyte measurement?

27. Ask your instructor to provide you a copy of a standard method involving titration. Identify the purpose of each reagent. Provide a rationale for each procedural step in the method.

28. You have been assigned the task of determining substance Z. Z does not react with any of the primary standards. Z does react with Y, but Y is not a primary standard. If you know that Y reacts with primary standard NaCl, explain how you could "standardize" substance Y so that it could be a "secondary standard" for the analysis of Z by titration.

Key Questions

1.   Diagram the process for Labs A and C described above following the pattern established below for Lab B.

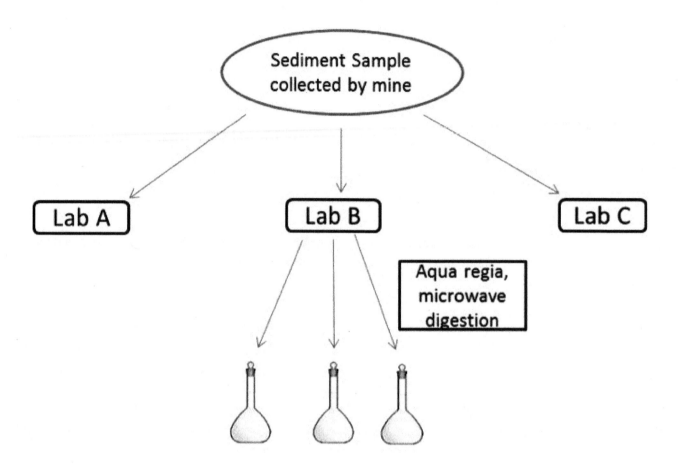

Consider this…

Table 1	Copper concentrations in the sediment as determined by each of the analysts.			
	Sediment Sample Mass (g)	**Copper concentration in digestate[a] (µg/mL)**	**Mass copper in digestate (µg)**	**Copper concentration in soil (µg/g)**
Lab A	Digested samples are clear with a layer of black and white particles on the bottom of the 25-mL volumetric flask.			
Sample 1	0.9838	0.134	3.35	3.41
Sample 2	1.0049	0.245		
Sample 3	1.0983	0.198		
			Average concentration	**4.67**
			Standard Deviation	
Lab B	Digested samples are clear, with only a few colorless to white particles on the bottom of the 25-mL volumetric flask.			
Sample 1	0.9794	0.765		
Sample 2	0.9486	0.759		20.00
Sample 3	1.0273	0.750		
			Average concentration	**19.27**
			Standard Deviation	
Lab C	Digested samples are clear with only a few colorless to white particles on the bottom of the 25-mL volumetric flask.			
Sample 1	0.9845	0.726		
Sample 2	0.9623	0.733		
Sample 3	0.9816	0.728		
			Average concentration	**18.67**
			Standard Deviation	

[a]Digestate is defined as the solution resulting after preparing the sample for analysis.

Key Questions

2. What is the final volume of each of the samples prior to GF-AAS analysis?

3. Notice that the mass of copper in the digestate for Lab A, Sample 1 has been calculated for you. Determine the equation used to calculate the mass of copper in the digestate and write it here.

4. Have one member of the group calculate the mass (µg) of copper in each 25 mL digestate solution for Lab A, another member calculate the mass of copper in each 25 mL digestate solution for Lab B, and a third calculate the mass for Lab C. Write these numbers in the appropriate spaces in Table 1. Compare the results for each lab. What do you notice?

5. Notice that the concentration of copper in the sediment has been calculated for Lab B, Sample 2. Determine the equation used to calculate the concentration of copper in the sediment from the mass of copper in the digestate and write it here.

6. Have each member of your group take one of the labs, calculate the copper concentration in the soil for each sample and write it in the appropriate location in Table 1.

7. Calculate the averages for each analyst. Do they match the given averages?

8. Calculate the standard deviations and write them in the appropriate box in Table 1.

9. Write a general equation to calculate the concentration of copper in each sample from the concentration in the digestate.

10. The three labs used two different sample preparation methods. List at least two differences between the methods.

11. Individually, examine the copper concentrations and standard deviations each lab found in their samples. How do the sets differ with respect to a) the analyte concentration and b) the precision?

12. Do any of the differences noted in your answer to Q10 provide evidence for the results you see in your answer to Q11? Explain.

13. Based on your answer to Q12 summarize the important considerations when preparing samples for analysis?

14. In your group, brainstorm some experiments you could use to determine if a sample is completely digested. Think about how you can ascertain the portion of copper retrieved from the sample. As a group, propose a scheme for testing the completeness of a digestion.

Consider this...

All labs tested the accuracy of the method by analyzing a Standard Reference Material (SRM). SRMs are real samples that contain concentrations of many analytes which are so well characterized that they are certified by the National Institute of Standards and Testing. SRMs are available in almost any imaginable sample matrix (sea water, river water, estuarine water, bituminous and anthracite coal, various types of sediment, etc.). The sediment SRM used by the analysts is certified at 4.63 ± 0.04 µg/g copper in the solid sample. All three labs followed the same procedure for the SRM analysis as they did for their samples. Analysis for copper by GF-AAS yields the following results:

Table 2 *Concentration of copper (µg/g) in the sediment SRM as determined by the analysts.*

	Lab A	Lab B	Lab C
Trial 1	1.36	4.95	4.59
Trial 2	1.42	4.88	4.64
Trial 3	1.31	5.02	4.61
Mean			
Standard Dev.			
% Recovery			

15. What concentration of copper should the analysts detect?

16. Calculate the mean and standard deviation of the analysts' results and place the numbers in the appropriate spaces on the table.

Percent Recovery is the amount of the analyte detected compared to the theoretical amount present: concentration of analyte detected / theoretical analyte concentration x 100%

17. Calculate the percent recovery for each lab and place the numbers in the appropriate spaces on the table.

18. As a group, discuss the results in Table 2. Summarize your observations.

19. Based on the results presented in Table 2, what can you say about the digestion process used by Lab A?

20. Someone suggests that the glassware in Lab B must be contaminated. Support or refute this statement with evidence from the data above.

Consider this...

Fortunately the analysts thought ahead and prepared several different blanks:

Method blank – All reagents were added to the digestion bomb with no sediment and analysis was carried out identical to samples.

Reagent blank – 10 mL of aqua regia was added to a 25-mL volumetric flask and diluted to the mark with deionized water and analyzed as a sample.

Instrument blank – Plain deionized water is analyzed as a sample.

Table 3 *Results of blank analyses.*

	Components	Copper concentration (µg/mL)		
		Lab A	Lab B	Lab C
Instrument Blank		0.001	0.002	0.002
Reagent Blank		0.001	0.036	0.001
Method Blank		0.003	0.034	0.002

21. Using the information provided, identify the components in each blank. Write your answers in the spaces provided in Table 3.

22. As a group, discuss the results. Record your thoughts. Does Lab A appear to have a contamination problem? Lab B? Lab C?

23. What is the source of the contamination? As a group, discuss the source of the contamination and propose a solution to the problem.

24. If the copper concentration for Lab B for the reagent blank had been very close to zero, what do you think would be the source of the contamination? Explain your reasoning.

25. Look back at your answer to Question 18. With the additional evidence provided in Table 3 do you still agree with what you wrote? If not, how would you modify your answer?

26. As a group, list three things you could do to assure minimal contamination in the sample prep and analysis.

27. You are the mine company manager. Which lab will you use for future analyses and why?

> Analysts need to assure the purity of all reagents, including distilled water, and the cleanliness of all equipment used. This is especially important when performing trace analysis. In addition to the blanks prepared in the lab, field blanks must be collected to assure that contamination is not picked up in the sampling process.

28. Brainstorm: How do you think a field blank is collected and treated? How is this different from a method blank?

29. List two ways this activity has increased your understanding of sample preparation.

30. List two tests you can use to determine whether or not a data set is reliable.

Application

31. You have been hired by an analytical firm to set up the sample processing procedure for mercury in several large (greater than 10 kg) fish. List and explain several things you must consider when carrying out the digestion process.

32. Often it is desirable to extract the analyte from the matrix before analyzing. Extraction removes many compounds that may interfere with the analysis. One extraction technique useful for analysis of trace concentrations of metals in water is co-precipitation with some sort of chelating agent like Ammoniumpyro-lidine-dithiocarbamate (APDC). The APDC forms an insoluble complex in water with metals. The precipitate is removed from the solution by filtration. The complex can then be dissolved in concentrated nitric acid with heat and diluted to a known volume with distilled water. Copper is extracted from 200.0 mL of riverine water by complexation with APDC, and the resulting precipitate is dissolved and diluted to 5.00 ml. The digestate is analyzed and found to contain 4.39 µg/ml Cu. What is the copper concentration in the river water? Locate information regarding ambient copper concentrations in natural systems and contamination levels. What can you say about the concentration of copper in this sample?

33. You have to prepare leaf samples for analysis. It is important that you dissolve as much of the leaf material as possible to extract the analyte. Propose a digestion process. Explain why you chose this method. (You may want to use your text to help you answer this question.)

34. SRMs are materials such as water, soil, plastics, blood, tissues, which are certified by the National Institute of Standards and Technology to contain specific concentrations of many analytes. Given that SRMs are real samples, how are the analyte concentrations known with such assurance?

Tinto, Rio; Kennecott Utah Copper website, http://www.kennecott.com/ 2011, accessed 5/20/11.

Environmental Literacy Council (ELC), Copper (2008), http://www.enviroliteracy.org/article.php/1029.html, accessed 5/20/11

Irwin, R.; van Mouwerik, M.; Stevens, L.;Seese, M.D.; and Basham, W., Environmental Contaminants Encyclopedia, Copper Entry, National Park Service, 1997, http://www.nature.nps.gov/hazardssafety/toxic/copper.pdf, accessed 5/20/11.

http://yosemite.epa.gov/opa/admpress.nsf/0/99138DEB5939D66D8525784D0076C9E0, accessed 6/2/2011.

Sample Preparation

Learning Objectives

Students should be able to:

Content

- Evaluate sample preparation methods for completeness.

- Determine the concentration of analyte in the sample from the concentration in the digestate.

- Explain how to assure sample integrity with respect to contamination and accuracy.

Process

- Form an understanding of the accuracy of a measurement. (Problem Solving, Critical thinking)

- Analyze methods for robustness and sources of error, and overall quality. (Critical thinking)

Prior knowledge

- Facility with solution dilution calculations.

- Ability to calculate standard deviation.

Further Reading

- D.C. Harris, *Quantitative Chemical Analysis*, 7th Edition, 2007 W.H. Freeman: USA, Chapter 28, pp. 644-659.

Author

Anne Falke

Consider this…

Copper mining has been an important industry in the United States for the past two centuries and remains so today. The Anaconda Copper Mine, located in Montana, provided much of the country's copper from 1892-1973 (EPA, 2011). Until 1915, the waste from the mining process, which contains substantial amounts of copper, was dumped into surrounding rivers. Copper is listed by EPA as a "priority pollutant" (Irwin, et al., 1997). It tends to bind strongly to sediments in the form of copper sulfide. Normal concentrations in soil range from 5-70 mg/kg (ELC, 2008). In 2003 the EPA found elevated concentrations of copper in the Missouri River sediments around the Anaconda mine (EPA, 2011). The site was declared a Superfund site in March of 2011.

In a hypothetical analysis of the sediments downstream from the mine, three different analytical labs were contracted to determine the copper content in sediments. A sediment sample was collected, mixed, divided into three portions, one of which was sent to each of the analytical labs.

The analyst from Lab A took three samples of about 1g each from the batch of sediment, placed them in three different vials, added 10 mL HCl to each vial, and heated them for 2 hours. The resulting solutions were decanted from each vial, transferred to three 25 mL volumetric flasks and diluted to the mark with distilled water.

The analyst from Lab B took the batch of sediment, ground it into fine particles with a mortar and pestle, and then placed three samples of about 1g each into Teflon bombs (a thick-walled Teflon container with a tight-fitting cover that can withstand high pressure) for digestion. The analyst then added 10 mL of aqua regia to each sample. Aqua regia is a mixture of HCl and HNO_3. The two acids together are more corrosive than either alone. The analyst sealed the bombs and microwaved them for 10 minutes. Microwaving samples in Teflon bombs allows for high pressure, high temperature digestion in a short amount of time. The resulting solutions were transferred to three 25 ml volumetric flasks and diluted to the mark with distilled water.

The analyst from Lab C also prepared three samples using the aqua regia bomb digestion following the same process as the analyst from Lab B.

All analysts use graphite furnace atomic absorption spectrometry (GF-AAS) to analyze the samples. GF-AAS is a technique especially well-suited for trace analysis of metals in aqueous samples.

Key Questions

1. Diagram the process for Labs A and C described above following the pattern established below for Lab B.

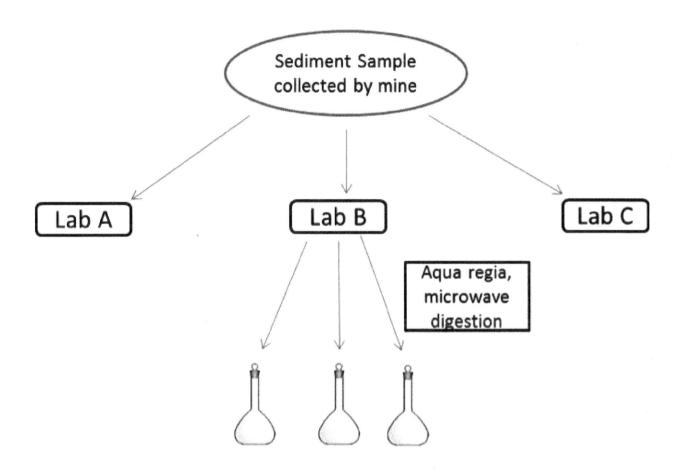

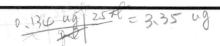

25 (0.9838g

0.9838g   0.134 ug / 25 mL = 3.35 ug

Consider this…

Table 1	Copper concentrations in the sediment as determined by each of the analysts.			
	Sediment Sample Mass (g)	**Copper concentration in digestate[a] (µg/mL)**	**Mass copper in digestate (µg)**	**Copper concentration in soil (µg/g)**
Lab A	Digested samples are clear with a layer of black and white particles on the bottom of the 25-mL volumetric flask.			
Sample 1	0.9838	0.134 ×25 mL	3.35	3.41
Sample 2	1.0049	0.245 ×25 mL	6.125	6.10
Sample 3	1.0983	0.198 ×25 mL	4.95	4.51
			Average concentration	**4.67**
			Standard Deviation	1.35
Lab B	Digested samples are clear, with only a few colorless to white particles on the bottom of the 25-mL volumetric flask.			
Sample 1	0.9794	0.765 ×25	19.13	19.53
Sample 2	0.9486	0.759	18.98	20.00
Sample 3	1.0273	0.750	18.75	18.25
			Average concentration	**19.27**
			Standard Deviation	0.90
Lab C	Digested samples are clear with only a few colorless to white particles on the bottom of the 25-mL volumetric flask.			
Sample 1	0.9845	0.726	18.15	18.44
Sample 2	0.9623	0.733	18.33	19.04
Sample 3	0.9816	0.728	18.20	18.54
			Average concentration	**18.67**
			Standard Deviation	0.32

6.125
1.0049

[a]Digestate is defined as the solution resulting after preparing the sample for analysis.

2. What is the final volume of each of the samples prior to GF-AAS analysis?

   25 ml

3. Notice that the mass of copper in the digestate for Lab A, Sample 1 has been calculated for you. Determine the equation used to calculate the mass of copper in the digestate and write it here.

   Each sample has 25 ml volume of solution, so in sample 1, 1 ml has 0.134 copper concentration × 25, it is 3.35 ug

4. Have one member of the group calculate the mass (µg) of copper in each 25 mL digestate solution for Lab A, another member calculate the mass of copper in each 25 mL digestate solution for Lab B, and a third calculate the mass for Lab C. Write these numbers in the appropriate spaces in Table 1. Compare the results for each lab. What do you notice?

   LAb A is very diffent then Lab B & Lab C

5. Notice that the concentration of copper in the sediment has been calculated for Lab B, Sample 2. Determine the equation used to calculate the concentration of copper in the sediment from the mass of copper in the digestate and write it here.

   Becaue in LaB B, Sample 2, 1 ml of volume of solution, contains 0.759 ug/ml Cu concentra, so mutiby 25 ml is 18.98 ug of mass Cu, so we use 18.98 ca in mas to divide sediment mass, then we get Cu concentra,

6. Have each member of your group take one of the labs, calculate the copper concentration in the soil for each sample and write it in the appropriate location in Table 1.

7. Calculate the averages for each analyst. Do they match the given averages?

   Y

8. Calculate the standard deviations and write them in the appropriate box in Table 1.

   Escel

9. Write a general equation to calculate the concentration of copper in each sample from the concentration in the digestate.

10. The three labs used two different sample preparation methods. List at least two differences between the methods.

11. Individually, examine the copper concentrations and standard deviations each lab found in their samples. How do the sets differ with respect to a) the analyte concentration and b) the precision?

12. Do any of the differences noted in your answer to Q10 provide evidence for the results you see in your answer to Q11? Explain.

13. Based on your answer to Q12 summarize the important considerations when preparing samples for analysis?

14. In your group, brainstorm some experiments you could use to determine if a sample is completely digested. Think about how you can ascertain the portion of copper retrieved from the sample. As a group, propose a scheme for testing the completeness of a digestion.

Consider this...

analyte = ion, molecules

All labs tested the accuracy of the method by analyzing a Standard Reference Material (SRM). SRMs are real samples that contain concentrations of many analytes which are so well characterized that they are certified by the National Institute of Standards and Testing. SRMs are available in almost any imaginable sample matrix (sea water, river water, estuarine water, bituminous and anthracite coal, various types of sediment, etc.). The sediment SRM used by the analysts is certified at 4.63 ± 0.04 µg/g copper in the solid sample. All three labs followed the same procedure for the SRM analysis as they did for their samples. Analysis for copper by GF-AAS yields the following results: *analyte = the thing we try to detect in our sample.*

Table 2 Concentration of copper (µg/g) in the sediment SRM as determined by the analysts.			
	Lab A	**Lab B**	**Lab C**
Trial 1	1.36	4.95	4.59
Trial 2	1.42	4.88	4.64
Trial 3	1.31	5.02	4.61
Mean *(average)*	*1.36*	*4.95*	*4.61*
Standard Dev.	*0.055*	*0.07*	*0.03*
% Recovery	*29.44 = 29%*	*106.91 = 107%*	*0.996 = 100%*

total

$$\frac{actual\ recovery}{themical\ recovery} \times 100\% = \boxed{}$$

$$\frac{4.95}{4.63} \times 100\% = 106.91\%$$

15. What concentration of copper should the analysts detect?

analyte   sample matrix

16. Calculate the mean and standard deviation of the analysts' results and place the numbers in the appropriate spaces on the table.

Percent Recovery is the amount of the analyte detected compared to the theoretical amount present:
concentration of analyte detected / theoretical analyte concentration x 100%

---

17. Calculate the percent recovery for each lab and place the numbers in the appropriate spaces on the table.

18. As a group, discuss the results in Table 2. Summarize your observations.

19. Based on the results presented in Table 2, what can you say about the digestion process used by Lab A?

20. Someone suggests that the glassware in Lab B must be contaminated. Support or refute this statement with evidence from the data above.

Consider this...

Fortunately the analysts thought ahead and prepared several different blanks:

Negative control (sample preparation)

Method blank – All reagents were added to the digestion bomb with no sediment and analysis was carried out identical to samples.

Reagent blank – 10 mL of aqua regia was added to a 25-mL volumetric flask and diluted to the mark with deionized water and analyzed as a sample.

Instrument blank – Plain deionized water is analyzed as a sample.

Table 3 *Results of blank analyses.*

	Components	Copper concentration (µg/mL)		
		Lab A	Lab B	Lab C
Instrument Blank	*water*	0.001	0.002	0.002
Reagent Blank	*aqua regia*	0.001	0.036	0.001
Method Blank	*water + reagent + solution*	0.003	0.034	0.002

21. Using the information provided, identify the components in each blank. Write your answers in the spaces provided in Table 3.

22. As a group, discuss the results. Record your thoughts. Does Lab A appear to have a contamination problem? Lab B? Lab C?

23. What is the source of the contamination? As a group, discuss the source of the contamination and propose a solution to the problem.

24. If the copper concentration for Lab B for the reagent blank had been very close to zero, what do you think would be the source of the contamination? Explain your reasoning.

25. Look back at your answer to Question 18. With the additional evidence provided in Table 3 do you still agree with what you wrote? If not, how would you modify your answer?

26. As a group, list three things you could do to assure minimal contamination in the sample prep and analysis.

27. You are the mine company manager. Which lab will you use for future analyses and why?

Analysts need to assure the purity of all reagents, including distilled water, and the cleanliness of all equipment used. This is especially important when performing trace analysis. In addition to the blanks prepared in the lab, field blanks must be collected to assure that contamination is not picked up in the sampling process.

28. Brainstorm: How do you think a field blank is collected and treated? How is this different from a method blank?

29. List two ways this activity has increased your understanding of sample preparation.

30. List two tests you can use to determine whether or not a data set is reliable.

Application

31. You have been hired by an analytical firm to set up the sample processing procedure for mercury in several large (greater than 10 kg) fish. List and explain several things you must consider when carrying out the digestion process.

32. Often it is desirable to extract the analyte from the matrix before analyzing. Extraction removes many compounds that may interfere with the analysis. One extraction technique useful for analysis of trace concentrations of metals in water is co-precipitation with some sort of chelating agent like Ammoniumpyro-lidine-dithiocarbamate (APDC). The APDC forms an insoluble complex in water with metals. The precipitate is removed from the solution by filtration. The complex can then be dissolved in concentrated nitric acid with heat and diluted to a known volume with distilled water. Copper is extracted from 200.0 mL of riverine water by complexation with APDC, and the resulting precipitate is dissolved and diluted to 5.00 ml. The digestate is analyzed and found to contain 4.39 µg/ml Cu. What is the copper concentration in the river water? Locate information regarding ambient copper concentrations in natural systems and contamination levels. What can you say about the concentration of copper in this sample?

33. You have to prepare leaf samples for analysis. It is important that you dissolve as much of the leaf material as possible to extract the analyte. Propose a digestion process. Explain why you chose this method. (You may want to use your text to help you answer this question.)

34. SRMs are materials such as water, soil, plastics, blood, tissues, which are certified by the National Institute of Standards and Technology to contain specific concentrations of many analytes. Given that SRMs are real samples, how are the analyte concentrations known with such assurance?

Tinto, Rio; Kennecott Utah Copper website, http://www.kennecott.com/ 2011, accessed 5/20/11.

Environmental Literacy Council (ELC), Copper (2008), http://www.enviroliteracy.org/article.php/1029.html, accessed 5/20/11

Irwin, R.; van Mouwerik, M.; Stevens, L.;Seese, M.D.; and Basham, W., Environmental Contaminants Encyclopedia, Copper Entry, National Park Service, 1997, http://www.nature.nps.gov/hazardssafety/toxic/copper.pdf, accessed 5/20/11.

http://yosemite.epa.gov/opa/admpress.nsf/0/99138DEB5939D66D8525784D0076C9E0, accessed 6/2/2011.

Instrumental Calibration (校对)

the action or process of calibrating an instrument or experimental reading

Learning Objectives

Students should be able to:

Content

- Use a correlation coefficient to ascertain the fit of data to a calibration curve

- Determine the minimum detectable concentration and lower limit of quantitation and distinguish between these.

- Determine the linear dynamic range from a set of calibration data.

Process

- Interpret tabulated data. (Information processing)

- Produce an xy graph of data. (Graphing)

- Interpret calibration curves. (Information processing)

Prior knowledge

- How to produce and interpret calibration curves.

- The meanings of linear regression, correlation coefficient, accuracy and precision.

Further Reading

- Harris, Daniel C. 2007. "Quality Assurance and Calibration Methods." In Quantitative Chemical Analysis, 7th ed., pp.78-87. New York: W.H. Freeman.

- Skoog, Douglas A., F. James Holler, and Stanley R. Crouch. 2007. "Introduction." Principles of Instrumental Analysis, 6th ed., pp.11-22. Belmont, CA: Thomson Brooks/Cole.

Authors

Anne Falke and Shirley Fischer-Drowos

PGLANA004001a POGIL

Consider this...

[handwritten: look, molecules]

In the real world, you really don't know the concentration of analyte in your samples. There are tools analytical chemists use to ascertain the precision and accuracy of their methods. Generally, you need to understand the chemistry of the analyte and the matrix.

Consider the following analysis of silver in drinking water samples. The researcher starts out determining the limitations of the method by analyzing a wide range of calibration standards. The results are below:

[handwritten annotations: ↑ X-asix   ↑ Y-asix   Y-asi   X-asix (we control)   we control x(line]

Table 1	
Ag (ppb)	**Signal**
0.00	0.000
2.00	0.013
10.0	0.076
20.0	0.162
30.0	0.245
50.0	0.405
75.0	0.598
100.0	0.748
125.0	0.817
150.0	0.872

Key Questions

1. Looking at the data, what do you notice about the relationship between the Ag concentration and the instrumental signal?

   [handwritten: the Ag concentration hit higher & more molecules, the instrumental signal curve is higher. In another words, the Ag concentration increases so does the instrument signal response more.]

2. Have one member of the group plot the signal (y-axis) vs. concentration (x-axis) on an Excel spreadsheet. Describe the pattern. _[Graphs may be supplied by your instructor. If graphs are supplied, you do not have to plot the data just look at Figure 1]_

   [handwritten: from 0 ~ 100 ppb seem to have linear. torwad stregthen, after it fell off the linear]

3. Using Excel, insert a "best fit" line, often called a trendline in spreadsheet programs. Highlight the linear trendline. While you have that window open, check "Display R-squared value on chart" (R^2 or correlation coefficient is a measure of how well the data points fit the trendline. The closer R^2 is to 1 the better the fit). What can you say about the fit of the data to the trendline? *[Again, this question refers to Figure 1 on supplied graphs]*

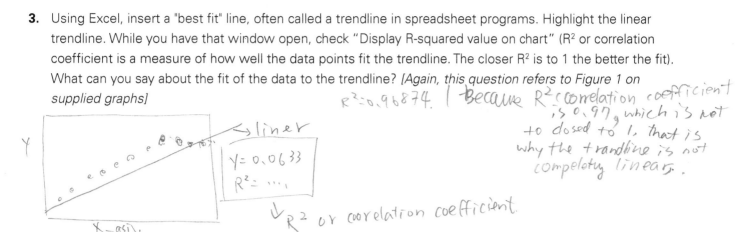

$R^2 = 0.96874$. Because R^2 (correlation coefficient is 0.97, which is not to closed to 1, that is why the trandline is not compelaty linear.

y $\leftarrow$ liner

$y = 0.0633$
$R^2 = \cdots$

$\downarrow R^2$ or correlation coefficient

X-asiy

4. As a group discuss how well the data fits the trendline. Do you notice some pattern to the points that do not fit? Describe the pattern.

When we see tho points starts to fall off (or flat), it means the detection is in maxium.

$\Rightarrow$ The middle between 75~100 is maxiam, after that is fall of or become flat, not linear, It that means, tho dectection in in maxium.

> In many types of instrumental measurements, instrumental signal is no longer linear at high analyte concentrations. This may be due to chemical effects or to the limitations of the instrument.

---

5. What leads you to think that the data presented above is not linear at high concentrations?

At this point, the analyst would examine the data and determine at what point the data begins to deviate from linearity. Divide the work among group members. Have one member plot concentrations from 0 – 125.0 ppb, the second plot 0 – 100.0 ppb, the third from 0 – 75.0 ppb and so on until there are six different graphs for this data. Be sure to include trendlines and correlation coefficients. Format the data labels so that you have at least 5 decimal places in the correlation coefficient and trendline equation. Regroup and compare your graphs. [If graphs are supplied refer to Figures 2-6].

6. Looking at all of the graphs, what does the correlation coefficient tell you about the linearity of the data?

7. What happens to the correlation coefficient as data points at higher concentrations are removed?

It is preferable to work in the region of the curve where the analyte signal is directly proportional to the analyte concentration. Analysts generally set a minimum acceptable correlation coefficient for a calibration. This varies with instrument and regulatory agency. Typical minimum acceptable correlation coefficients range from 0.995 to 0.9999.

8. As a group, decide which plot provides the best fit of the data without severely limiting concentration range over which the calibration can be used. Explain your answer.

9. What is the correlation coefficient for the selected graph?

10. Look at the graph with the correlation coefficient just over 0.995. What do you notice about the last datapoint? Does this seem to be a result of random error or is it the start of rollover (negative deviation from linearity)?

11. a. Have a member of your group calculate the concentration of a sample with an analytical signal of 0.532 on the curve covering 0 – 100 ppb. Have a second member calculate the concentration of a sample with an analytical signal of 0.532 on the curve covering 0 – 75 ppb. As a group, evaluate how these numbers compare. What is the percentage error introduced using the curve with a hint of rollover?

b. Discuss the error with your group. Do you think it is acceptable? Why or why not?

12. Based on the answers to the previous questions, decide as a group which is the best curve to use for this analysis. Explain your answer.

13. Based on the curve your group decided upon, what is the highest concentration standard that is still has a reasonably linear signal?

The value determined above is the upper limit of analysis. Any sample with a concentration greater than the upper limit cannot be reliably analyzed using the established calibration curve. For this reason it is desirable to choose the upper limit so that the curve is acceptably linear, but the available range is not excessively limited.

Consider this...

To continue establishing instrumental limitations the analyst carries out the following analysis by graphite furnace atomic absorption spectrophotometry. Note that the calibration is identical to the one established in the first part of this activity.

Table 2	
Ag (ppb)	Signal
Calibration Standards	
0.00	0.000
2.00	0.013
10.0	0.076
20.0	0.162
30.0	0.245
50.0	0.405
75.0	0.598
100.0	0.748
Detection Limit Analysis	
0.00	0.003
0.00	0.007
0.00	0.004
0.00	0.006
0.00	0.008
0.00	0.004
0.00	0.005
2.00	0.012
2.00	0.017
2.00	0.015
2.00	0.013
2.00	0.019
2.00	0.017
2.00	0.016

Notice the information that follows the calibration curve data. The analyst has analyzed 7 replicate blank samples (no silver in the sample) and 7 replicate samples with a low concentration (1-5 times the detection limit) of silver. This is done to determine the minimum concentration of analyte that may be detected (analyte is present, but not necessarily at a concentration great enough to provide data reliable enough to quantify) and the minimum concentration that is quantifiable (present in a high enough concentration to measure with reasonable accuracy).

14. Discuss the difference between detection and quantitation with your group members. Write a sentence or two identifying that difference.

15. Which should be the larger number? Why?

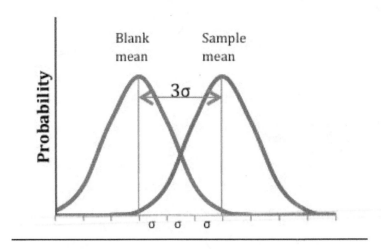

The variation in replicate analysis is a function of the randomness of the instrumental signal (noise). The analyst must be able to separate a real signal from the noise. Therefore, increased noise increases the limits of detection and quantitation.

The figure above represents the distribution of blank data and sample data. Note that the sample mean is 3 standard deviations 3σ greater than the blank mean. About 1% of the distribution of the blank data is greater than the sample mean. The minimum detectable signal is defined as the average signal for the blank + 3 times the standard deviation (s) of the low level sample signals. Half of the distribution for the sample data lies below the sample mean. What this means is that there is about a 1% chance that a blank will be detected as a real sample and a 50% chance that a sample with a mean at the detection limit will be viewed as a blank.

16. Calculate the average signal for the replicate blank analysis.

17. Calculate the standard deviation for the replicate 2.00 ppb standards.

18. Calculate the minimum detectable signal for the data above.

19. Using the calibration curve chosen in question 12, individually calculate the concentration that corresponds to the minimum detectable signal. Compare your answer with your group members. When you have a single value that you all agree on write it here.

The concentration that corresponds to the minimum detectable signal is defined as the lower limit of detection (LLOD) (also called minimum detectable concentration or detection limit). *If the intercept for the calibration curve and the average blank signal are both approximately zero* as is typically the case in atomic absorption spectrophotometry, the LLOQ can be estimated from 3 times the standard deviation of the low concentration signals divided by the slope of the calibration curve.

$$LLOD = 3s/m$$

---

20. Calculate the LLOD using 3s/m method to prove that the two calculations yield approximately the same concentration (of the same order of magnitude).

21. What are the units for the LLOD?

22. As a group, summarize the calculations for detection limits:

    a. Minimum Detectable Signal

    b. Lower Limit of Detection (exact method)

    c. Lower Limit of Detection (estimated method)

23. In what situations would you use each of the above?

The lower limit of quantitation (LLOQ) is defined as the concentration which corresponds to the signal that is 10 times greater than the noise.

---

24. Write an equation to determine the LLOQ.

25. What is the signal for the LLOQ for the data set above?

26. Using the calibration curve established in Q12 determine the concentration corresponding to the LLOQ.

As with the LLOD, for methods that produce a calibration with an intercept of approximately zero, the LLOQ can be estimated:

$$LLOQ = 10s/m$$

27. Identify the assumptions used in the estimation LLOQ calculation.

28. As a group, compile a clear set of directions for your lab technician describing how to determine the LLOD and LLOQ for a new instrument.

The linear range is defined as the range over which the instrumental signal is proportional to the concentration of the analyte. The lower end of the linear range is the LLOQ and the upper limit is the highest concentration standard for which the signal is still linear. The linear range is expressed in concentration terms.

29. What is the linear range (upper and lower limit) for the set of data in Table 2?

30. Since the linear range encompasses the analyte concentrations for which the signal is linear, a signal out of that range cannot be reliably quantified. What can you do to reliably analyze a sample that yields a signal of 0.936 on this curve?

31. What can you do to get reasonable quantitation of a sample that initially yields a response of 0.010 with this curve?

32. List and explain two ways that your understanding of the principles and use of calibration curves has improved.

33. How does graphing tabulated data enhance your ability to interpret it?

Application

34. Given the following set of emission data, determine linear range of quantitation for the analysis. Calculate the LLOD using both the exact and estimated methods and compare to the lower limit of your linear quantitation range.

[Pb] (ppb)	Emission counts
0	25
0.005	918
0.010	1809
0.025	4987
0.050	10143
0.075	15203
0.100	19357
0.200	39362
0.300	48587
0.400	55072
0.500	61093
0	62
0	115
0	89
0	10
0	53
0	121
0	105
0.005	928
0.005	881
0.005	985
0.005	1002
0.005	993
0.005	911
0.005	873

35. The following samples were analyzed for silver. To each digestion flask 200 mL aliquots of 10 different drinking water samples were added. Silver was extracted by a co-precipitation procedure and collected by filtration. The precipitate was digested in 200 µl hot concentrated nitric acid and diluted to 5 mL in a dilute phosphate solution. Standards were prepared according to the following table. The samples were analyzed by graphite furnace AAS and compared to the calibration curve. Results of the analyses follow:

Standards Ag (ppb)	Absorbance
0.00	0.000
2.00	0.013
10.0	0.076
20.0	0.162
30.0	0.245
50.0	0.405
75.0	0.598
100.0	0.748

Samples	Absorbance
08002	0.079
08003	0.058
08004	0.129
08005	0.082
08006	0.797
08007	0.063
08008	0.009
08009	0.028
08010	0.136

Calculate the concentration of silver in each sample. Are there any problems? How will you address each of the identified problems?

36. Identify the data you would need to gather to establish the linear range for a particular analysis.

Quality Assurance Measures

Learning Objectives

Students should be able to:

Content

- Calculate the spike recovery and relative deviation between replicate samples.

- Use spike recovery and relative deviation between replicate samples to assess and validate sample analysis quality.

- Use spike recovery and relative deviation between replicate samples to assure accuracy and precision of an analysis.

Process

- Validate data. (Problem solving)

- Interpret tabulated data. (Information processing)

Prior knowledge

- Evaluation of calibration curves including linear range and limit of detection, linear regression, correlation coefficient, accuracy and precision.

Further Reading

- Harris, D.C. 2007. *Quantitative Chemical Analysis*, 7th Edition, pp.78-87. New York: W.H. Freeman.

- Skoog, D.A., F.J. Holler, and S.R. Crouch. 2007. *Principles of Instrumental Analysis*, 6th Edition, p11-22. Belmont, CA: Thomson Brooks/Cole.

Authors

Anne Falke and Shirley Fischer-Drowos

PGLANA004006a POGIL

Consider this…

Consider the following analysis of silver in drinking water samples. To each of three digestion flasks, 200 mL aliquots of a drinking water sample were added. Silver was extracted by a co-precipitation procedure and collected by filtration. The precipitate was digested in 200 µl hot concentrated nitric acid and diluted to 5 mL in a dilute phosphate solution. Two of the samples were treated identically. They are called "sample" and "duplicate" in the list below. Five µL of a 10.00 ppm Ag solution was added to the third sample resulting in a 10.00 ppb spike. Standards were prepared according to the following table. The samples were analyzed by graphite furnace AAS. Results of the calibration and analysis follow:

Run number	Ag (ppb)	Absorbance
1	Calibration Standards Calibration Blank	0.000
2	2.00	0.013
3	10.0	0.076
4	20.0	0.162
5	30.0	0.245
6	50.0	0.405
7	75.0	0.598
8	100.0	0.748
9	Instrument Performance Check (30.0 ppb)	0.243
10	Calibration Blank	0.004
11	Sample	0.030
12	Duplicate Sample	0.032
13	Laboratory Fortified Sample Marix (1.0 ppb Ag added to sample)	0.109
14	Instrument Performance Check (30.0 ppb)	0.232
15	Calibration Blank	0.003

Table 1

Note that there are several analyses performed in addition to the calibration curve and sample. The additional analyses are Quality Assurance (QA) samples. These samples must be analyzed to assure the precision and accuracy of the analysis. The frequency of QA samples in an instrumental analysis is often specified by various regulatory agencies such as EPA and FDA to assure the integrity of analysis and is dependent on who is requesting the analysis. *standard method development*

Key Questions

1. What analyses have been performed in addition to the calibration curve? Discuss these and decide as a group why you think each of these analyses was performed. Provide an explanation for each additional analysis.

   (8) calibration blank → 0.004 → closed to ①.

   It means instrument pretty closed.

   when we look 9 ~ 13, then ⑭ is 0.232. It ⑭ is to make sure all instruments round the same. the answer is 5% different.

2. Plot the calibration curve, insert a trendline and determine the equation for the line.

   $y = 0.0077x + 0.0083$

   $R^2 = 0.9775$.

An Instrument Performance Check is one of the intermediate calibration standards. The value obtained must be converted to a concentration using the calibration data and then compared to the expected value. Generally, recovery of an Instrument Performance Check must be between 90 and 110% (some industries apply tighter constraints) before continuing with the analysis. If recovery falls outside this range, it means that the instrumental measurements are not under control.

---

3. Use the calibration curve to calculate the concentration of the Instrument Performance Check.

   $0.245 = 0.0077x + 0.0083$

   $0.0077x = 0.245 - 0.0083$

   $x = 30.740$

   $0.232 = 0.0077x + 0.0083$

   $0.0077x = 0.232 - 0.0083$

   $x = 29.05$.

4. Compare that value to the expected concentration (10.0 ppb.) Calculate the percent recovery (compared to the expected concentration.)

$$\frac{30.740}{30} \times 100 = 102.5 \qquad \frac{29.05}{30} \times 100 = 96.83.$$

5. Does analysis of the Instrument Performance Check provide information about accuracy or precision or both? Explain.

more accuracy because wat use instrument to define the result, but we could say little precision, because there is comparas between 9 & 14.

6. As a group, decide if the recovery of the Instrument Performance Check is acceptable. As if your group were reporting this information to your client, explain the rationale for your decision using complete sentences.

yes. it is acceptable.

> The Calibration Blank is analyzed to assure that there is no instrumental contamination. It is prepared identically to the other standards (i.e. same solvents, same process), but with no analyte added. The calculated concentration must be below the Minimum Detection Limit (MDL) to continue.

7. Brainstorm some potential sources of contamination.

8. If the MDL for this method is 0.1 ppb, is there evidence of any silver contamination in the instrument? Explain your reasoning.

Good Laboratory Practice (GLP) requires that an Instrument Performance Check and a Calibration Blank be analyzed after analysis of the calibration standards, but before any samples are analyzed, regularly throughout the analysis of samples (every 10 samples is common), and after analysis of all samples. If either the check or zero standard is out of control, protocol requires that the analyst redo the calibration curve and reanalyzes all samples analyzed between the previous check standard and the one deemed unacceptable.

---

9. Now look at the sample labeled "Duplicate Sample." The Duplicate Sample is identical to the sample in every way except that it is prepared and analyzed separately. Does the Duplicate Sample provide information about precision, accuracy, or both? Explain in complete sentences.

The frequency at which Duplicate Sample must be prepared and analyzed is designated by the regulations governing the particular industry. Generally, an analyst prepares one duplicate for every batch of 10 or 20 samples as specified. Relative deviation is defined as the difference between the calculated concentration of the original sample and the concentration of the duplicate sample divided by the original sample concentration. Multiply by 100 to get percent relative deviation.

$$\frac{|\, Sample\ concentration - duplicate\ concentration \,|}{sample\ concentration} \times 100\%$$

Generally, a percent relative deviation of less than 10% is acceptable. A deviation greater than 10% must be flagged and explained.

---

10. Calculate the concentration of the Sample and the Duplicate Sample. Calculate the percent relative deviation for this analysis. Is this acceptable? What does it tell you about the precision of this analytical technique with respect to silver?

11. Now, examine the Laboratory Fortified Sample Matrix (LFM). The LFM is prepared by adding a known amount of analyte to a particular sample. The LFM is analyzed to determine how much of the added analyte is recovered. Does this give you information about accuracy, precision, or both? Discuss this with your team members and explain in grammatically correct sentences.

12. A 10.0 ppb spike is added to the LFM. What is the concentration of silver detected in the LFM?

13. Accounting for the amount of silver detected in the sample, how much of the silver detected in the Laboratory Fortified Sample Matrix is due to the added silver?

14. Given what you know about the concentration of added silver, what percent of the added silver was detected? Explain how you determined the percent of added silver detected.

15. As a group, develop an equation for calculating percent recovery of an LFM in a similar situation with a given amount of analyte added to a sample with a known amount of the analyte.

Industry standards generally require that the recovery of the added analyte in a Laboratory Fortified Sample Matrix range from 75 – 110%. (Some industries and regulatory agencies set tighter conditions.)

16. Is the recovery of the added silver in the LFM acceptable? Explain.

17. Discuss the acceptable range for recovery of added analyte in the Laboratory Fortified Sample Matrix with your group. Why do you think there is more leeway on the lower end (75%) than on the upper end (110%)?

18. Complete the following table with the calculations for each QA parameter.

Parameter	Equation
Check Standard Recovery	
Percent Relative Deviation	
Spike Recovery	

19. As a group summarize what you learned in this activity.
    a. Write down at least three key concepts.
    b. Write at least one question that is lingering after completing this activity.

20. Look back at the table at the beginning of this activity. What features make it easy to make sense of the table? What features of the table would you change to make it easier to read? Do you think the table is in an appropriate format for a professional report or publication? Why or why not?

Applications

21. Analysis of lead by emission spectroscopy yields the following for the calibration standards, a sample, and the same sample spiked with 0.050 ppb lead. Calculate recovery of added lead in the Laboratory Fortified Sample Matrix.

[Pb] (ppb)	Emission counts
0	13
0.005	1018
0.010	2009
0.025	4987
0.050	10643
0.075	15403
0.100	19457
0.200	38062
Sample	1530
Laboratory Fortified Sample Matrix	9451

22. You are charged with analyzing 25 water samples for lead. EPA specifies action if more than 10% of the samples test at greater than 0.015 mg/L. Additionally, EPA mandates a QA protocol of one Sample Duplicate for every 10 samples and one Laboratory Fortified Sample Matrix for every 20 samples. The instrumental protocol calls for a calibration curve of at least four non-zero standards plus a Calibration Blank. The calibration curve must be verified before analyzing any samples, after every 10 analyses, and again at the end of the analysis. Prepare a table that lists the sequence in which the samples, spikes, duplicates and standards should be run in order to be compliant with the EPA regulations.

23. In addition to all of the quality assurance samples discussed in the activity, many regulatory agencies require analysis of a Standard Reference Material (SRM). SRMs are real samples that are well characterized. Analyte concentrations are known with a great deal of certainty. SRMs are available in almost any imaginable sample matrix (sea water, river water, estuarine water, bituminous and anthracite coal, various types of sediment, etc.). The National Institute of Standards and Technology (NIST) is a provider of SRMs. Analysis of a NIST drinking water standard certified at 19.73 + 0.02ppb silver gave the following results based on the calibration curve established from the data in Table 1.

Run #	Absorption
1	0.147
2	0.152
3	0.143
4	0.152
5	0.149
6	0.145

Based on what you have discovered in this activity, discuss whether these results are acceptable. Support your answer. If you were in the position of analyzing samples for silver concentration for a client, what would you tell that client? Be as thorough as possible.

24. In industry, time is money. Each analysis requires a specified amount of time. If each analytical run requires 2 minutes, what is the minimum amount of time required to sufficiently analyze a single sample? Consider all parts of the analysis from calibration to Quality Assurance. Explain where all of the time is used.

25. A few individuals in a small town thought there might be toxic concentrations of mercury in the drinking water. They filled an empty water bottle with water from the tap and brought it in to the local university to be analyzed. When the results came back, the concentration reported did indicate toxicity. The individuals decided to sue the town for knowingly poisoning the residents of the town. You are now employed by the town to handle the problem. What are you going to tell the lawyers? Consider everything covered in this activity.

Instrumental Calibration: Method of Standard Additions

Learning Objectives

Students should be able to:

Content

- Recognize when the method of standard additions is needed. (matrix effects)
- Distinguish between the method of standard additions and normal calibration using standards.
- Determine the concentration of an unknown in an original sample using the method of standard additions.

Process

- Interpret calibration graphs. (Information Processing)
- Calculate concentrations of unknown solutions from standard addition data. (Information Processing)
- Read and interpret a standard method. (Information Processing)

Prior knowledge

- Calibration using standards. (external and internal calibration)
- Use of statistical tests for outliers and comparing means.
- Linear regression, linearity, detection limit.
- Use of Microsoft Excel to plot data and obtain best fit lines or polynomials.
- Completion of *Quality Assurance Measures* activity.

Further Reading

- Creed, J.T., Martin, T.D., and O'Dell, J.W. Method 200.9, *Determination of Trace Elements by Stabilized Temperature Graphite Furnace Atomic Absorption*, Revision 2.2. US Environmental Progection Agency [online] Nov. 4, 2010.
  http://water.epa.gov/scitech/methods/cwa/bioindicators/upload/2007_07_10_methods_method_200_9.pdf
- Harris, D.C., 2010. *Quantitative Chemical Analysis*, 8th Edition, W.H. Freeman: USA, p.106-108, 330, and 494.
- Harvey, D.J. *Chem.* Ed. 2002, 79, pp.613-615.

Authors

Christine Dalton and Mary Walczak

PGLANA004007a

Section 1. Determining When the Method of Standard Additions is Needed

In this section we will perform a number of different quality assurance checks. For convenience, Appendix C contains an abbreviated list of the checks that will be conducted. You may wish to tear off Appendix C for use throughout the activity.

Consider this...

Determining the Linear Dynamic Range

The following data tables were collected for analysis of industrial waste samples for lead using a graphite furnace atomic absorption instrument according to EPA Method 200.9 *(Determination of Trace Elements by Stabilized Temperature Graphite Furnace Atomic Absorption)*. Table 1 shows the data for five solutions of lead at the same concentration (7.5 ppb) analyzed to demonstrate instrument stability. The % Relative Standard Deviation (RSD) must be less than 5% for the instrument stability check to pass. *Note: Appendix A includes an excerpt from*
Section 3 of the EPA Method with definitions of useful terms.

Table 1 *Solutions Prepared for Atomic Absorption Analysis of Industrial Waste Samples for Lead*

Demonstrate Instrument Stability (Section 11.4.3*)		
Solution Number	Concentration Pb Standard (ppb)	Absorbance
1	7.5	0.036
2	7.5	0.036
3	7.5	0.035
4	7.5	0.037
5	7.5	0.034

* The section number refers to the EPA Method 200.9

Key Questions

1. Solutions 1–5 are analyzed to demonstrate instrument stability. According to the method, "the resulting relative standard deviation (RSD) of absorbance signals must be <5%." What is the %RSD for these five solutions? Does the instrument pass the stability check? Indicate your decision on the Table in Appendix C. *(Note: We will collect all our quality assurance checks in one place in Appendix C.)*

$$\%RSD = \frac{Stdev}{Avarage} \times 100\%$$

2. The method says "instrument stability must be demonstrated by analyzing a standard solution with a concentration 20 times the Instrument Detection Limit (IDL) a minimum of five times." Solutions 1-5 are run to demonstrate instrument stability. What is the IDL for this instrument? Check your answer by conferring with another group. Resolve any discrepancies before continuing.

Consider this...

Table 2 below shows the calibration run for lead following EPA method 200.9 and the solutions used to determine the method detection limit (MDL). The plot for the data in Table 2 is shown on the next page along with the least-squares line and correlation coefficient (R^2) for different concentration ranges.

Table 2 *Solutions Prepared for Atomic Absorption Analysis of Industrial Waste Samples for Lead*

Calibration Run (Section 11.4.4)		
Solution Number	**Concentration Pb Standard (ppb)**	**Absorbance**
6	0	0.001
7	1.6	0.018
8	8	0.048
9	20	0.085
10	40	0.186
12	60	0.276
13	90	0.422
14	110	0.535
15	140	0.657
16	170	0.704
Method Detection Limit (Section 9.2.4)		
Solution Number	**Concentration Pb Standard (ppb)**	**Absorbance**
17	200	0.795
18	1.0	0.001
19	1.0	0.002
20	1.0	0.001
21	1.0	0.003
22	1.0	0.002
23	1.0	0.002
24	1.0	0.002

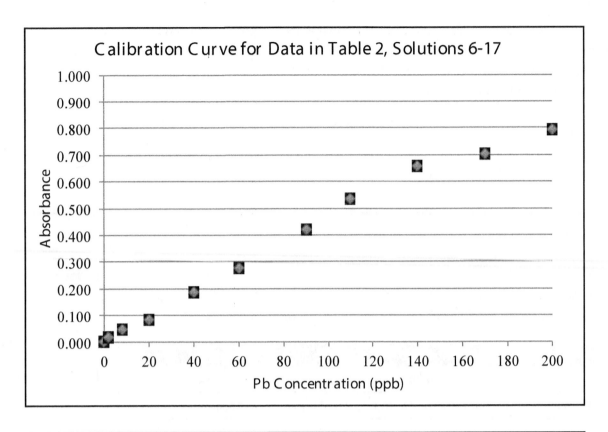

Calibration Curve for Data in Table 2, Solutions 6-17

range	equation	R²
0 to 110 ppb	y = 0.00474 x + 0.001	0.9979
0 to 140 ppb	y = 0.00471 x + 0.002	0.9988
0 to 170 ppb	y = 0.00440 x + 0.012	0.9907
0 to 200 ppb	y = 0.00416 x + 0.021	0.9868

Key Questions

3. What is the upper limit for Pb analysis using this method? How did you determine this limit?

4. The method detection limit (MDL) is determined once per year by "processing seven replicate samples through the entire method." These seven replicates are prepared "using reagent water (blank) fortified at a concentration of two to three times the estimated [IDL]." To calculate the MDL, the Student's t value for a confidence level of 99% and for n-1 degrees of freedom is multiplied by the standard deviation of the concentrations of the seven replicates. If t=3.14 for seven replicates, what is the MDL for this analysis using solutions 18-24 in Table 2? (Hint: We want the MDL in concentration units (ppb), so we must first change the absorbance values for solutions 18-24 into concentrations.)

5. Using the information in Q2-Q4, fill in the following table

IDL	
MDL	
Upper limit	

The linear dynamic range for an analysis is the lowest to highest concentrations that can be measured with the method. The lower limit of quantitation (LLOQ) is the smallest concentration that can be measured with reasonable accuracy. Based on the equations below, where s is the standard deviation of the concentrations of the solutions used to determine the MDL and m is the slope of the best calibration curve, the MDL can be used to calculate the LLOQ:

$$MDL = \frac{3*s}{m} \qquad LLOQ = \frac{10*s}{m}$$

$$\frac{s}{m} = \frac{MDL}{3} \qquad LLOQ = \frac{10*s}{m} = \frac{10*MDL}{3}$$

Using the information entered in the table above, determine the "linear dynamic range" for Pb analysis using this method. Compare your answer to that of another group.

Linear dynamic range	

Consider this...

Assessing Laboratory Performance

According to the EPA 200.9 method, assessing laboratory performance is mandatory on a daily basis. If any of the Quality Controls do not pass, then the sample data is invalid and the problem must be found and corrected prior to running samples. Solutions 25 through 48 in Table 3 include several solutions that are utilized for assessing laboratory performance along with analysis of several samples.

Table 3 Solutions Prepared for Atomic Absorption Analysis of Industrial Waste Samples for Lead		
Sample analysis (11.4) with Performance Assessment (9.3 and 9.4)		
Solution Number	Sample Name	Absorbance
25	Laboratory Reagent Blank	0.004
26	Calibration Blank	0.002
27	Instrument Performance Check (25 ppb)	0.123
28	Laboratory Fortified Blank (25 ppb)	0.122
29	Sample 1	0.408
30	Sample 2	0.521
31	Sample 3	0.446
32	Duplicate Sample 3	0.448
33	Sample 4	0.512
34	Sample 5	0.926
35	Sample 6	0.489
36	Laboratory Fortified Sample Matrix: Sample 6 + 25 ppb Spike	0.568
37	Sample 7	0.506
38	Sample8	0.579
39	Laboratory Reagent Blank	0.003
40	Calibration Blank	0.001
41	Instrument Performance Check (25 ppb)	0.112
42	Sample 9	0.482
43	Sample 10	0.009
44	Sample 11	0.819
45	Sample 12	0.486
46	Laboratory Reagent Blank	0.002
47	Calibration Blank	0.002
48	Instrument Performance Check (25 ppb)	0.125

Key Questions

Exploring the Solutions from Table 3

6. Referring to solutions numbered 1-48 in Tables 1-3, match all 48 solution numbers with the eight categories of solutions listed in column one of the table. Add the solution numbers to the second column of the table from Appendix C.

Solutions	Solutions from Tables 1-3
Instrument Stability Check	
Calibration Standards	
Method Detection Limit	
Laboratory Reagent Blank	
Calibration Blank	
Instrument Performance Check	
Laboratory Fortified Blank	
Samples	
Duplicate Samples	
Laboratory Fortified Sample	

Assessing blanks for contamination

7. According to the EPA method there are several types of "blanks" used in the analytical run. Solutions 25-48 in Table 3 include several of these blank solutions. Identify the different types of blanks in Table 3 and add these to the following table. Specify the solution numbers for each type from Table 3 and add those to the following table. List the purpose and composition of each type of blank in the run. Note: Appendix A includes an excerpt from Section 3 of the EPA Method with definitions of useful terms, including several types of blanks.

Solution	Solution Number(s)	Purpose of Solution	Composition of Solution	Requirement
			Contains all reagents in the same volumes as used in processing samples	[Analyte] < 2.2 x MDL
		Used to check that the instrument is zeroed correctly throughout the run		[Analyte] < MDL
	28			% Rec = 85-115%

8. Using the best least-squares line (determined in Q3) calculate the concentration for each laboratory reagent blank and calibration blank. Add the concentrations of these blanks to the last column of Table 3 in Appendix B. *Note: Do not add any other concentrations to this table until you are told to do so.*

9. Using the information in the table in Q7, determine if each laboratory reagent blank (LRB) and calibration blank in Table 3 meets the requirement for that type of blank to pass the contamination check. Recall that the MDL and IDL were determined in Q5. Do all of the tests pass for the LRBs and calibration blanks? If all of the LRBs and calibration blanks pass the test, then add a check in the "Pass?" column in the table in Appendix C. If they do not pass the test, then write "Fail" in the "Pass?" column.

Exploring samples and assessing sample duplicates

10. Table 3 includes several solutions labeled as "samples." How many different samples were analyzed? How many replicates of the samples were performed? According to the EPA method, a replicate must be analyzed for every 10 samples. Does this run meet this requirement?

11. What are the differences between solutions 31 and 32? What is the purpose for running both of these samples? If the method states that the difference between a sample and a sample duplicate must be less than 10%, does this run pass this criterion? Add a check or "Fail" to the table in Appendix C to designate if the duplicate passes the criterion for performance assessment.

Checking concentrations of known solutions

12. Calculate the concentration of the Laboratory Fortified Blank (LFB) in Table 3 and add the concentration to Appendix B. What is the percent recovery (i.e., calculated value/expected value x 100) for the LFB if the amount of added Pb is 25 ppb? Does this value fall within the acceptable limits? Add a check or "Fail" in the "Pass?" column in the table in Appendix C to specify passing or failing of the LFB.

13. The run sequence includes several Instrument Performance Checks as shown in Q7. What is the purpose of including these samples in the run?

14. Calculate the concentration of the instrument performance check (IPC) in Table 3 and add the concentration for each check to the table in Appendix B. The instrument performance check is done using a certified reference material with a known concentration of 25 ppb. Based on the calculated concentration of the standard reference material, determine the percent recovery of the reference material (i.e., calculated value/ expected value x 100). The EPA method states that the first IPC check should fall within 5% of the calibration, while later IPC checks can be within 10% of the calibration. Do the IPC check concentrations fall within the acceptable limits? Add a check or "Fail" in the "Pass?" column in the table in Appendix C to indicate if the IPC passes the performance check.

15. What are the differences between solutions 35 and 36? What is the purpose for running both of these samples? The EPA method states that the Laboratory Fortified Sample Matrix must be analyzed once for every ten samples, and it must be a duplicate of the aliquot used for sample analysis. Does sample 36 meet this criterion?

16. Calculate the concentrations of solutions 35 and 36 and add them to Table 3. For the Laboratory Fortified Sample Matrix to pass the performance assessment, the percent recovery of the analyte must be within 70-130%. This is calculated by subtracting the unfortified sample's concentration (solution 35) from the fortified sample's concentration (solution 36) and dividing by the concentration of analyte added to the sample and multiplying the result by 100. Does sample 36 pass the performance check? Add a check or "Fail" in the "Pass?" column in the table in Appendix C to indicate if the Laboratory Fortified Sample Matrix (LFM) passes the test.

17. In Q12 through Q16, solutions of known concentration were analyzed and the percent recovery was checked against a criterion of the method. What is the difference between the LFB, the IPC, and the LFM if each had 25 ppb Pb added?

18. If the LFM is the only solution of known concentration to fail the criterion for percent recovery, brainstorm reasons why this solution would not be within the acceptable range? Come to consensus.

19. Review the data you have accumulated in the Table from Appendix C. Do all parts of the assessment pass so that the analysis of the samples is acceptable? Explain how you came to your conclusion.

In chemical analysis, the presence of other analytes in the sample can interfere with the detection of the analyte of interest. This is especially true of natural samples, such as pond water. When this occurs it is called a "matrix effect" because the compounds in the sample interfere with analysis. In such cases, the method of standard additions can be used to eliminate differences between samples and calibration standard solutions.

20. What is the logical next step in this analysis?

Section 2. Determining Concentration from Standard Addition Data

Consider this...

Calibration standards were prepared using 100-mL volumetric flasks and a 75.0 ppm standard solution of analyte. The amount of standard added to each flask is listed in Table 4. Each flask was then diluted to volume with deionized water and thoroughly mixed. *Note that solution shading (see legend) is not meant to indicate the lack of mixing but just illustrate relative amounts.*

Table 4   Calibration Standards

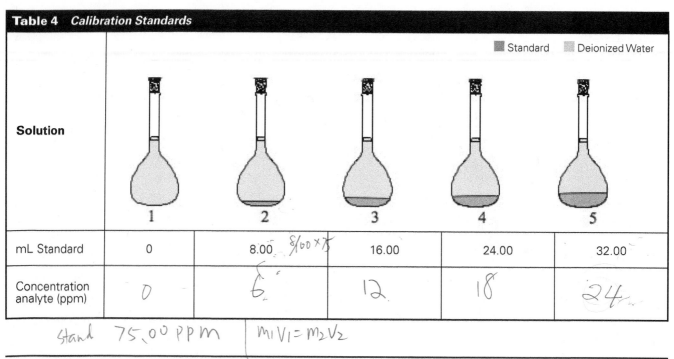

	1	2	3	4	5
Solution					
mL Standard	0	8.00 *8/60 × 75*	16.00	24.00	32.00
Concentration analyte (ppm)	0	6.	12	18	24

stand 75.00 ppm $M_1 V_1 = M_2 V_2$

Key Questions

21. What happens to the relative amounts of standard and water among the calibration standards?

When the ml stand going up, then water is going done. That mean the sample solution more concentration, less water.

22. As a group, consider the calibration standards samples in Table 4. First, devise a general method to calculate the concentration of analyte in a prepared solution.

Then divide solutions 1-5 among group members and calculate the concentrations. Add the answers to the Table and check that concentrations increase with the amount of standard used.

23. Suppose you wanted to make a 16.0 ppm calibration standard. Describe how you would prepare this solution.

Consider this...

A set of Standard Addition solutions were prepared using 100-mL volumetric flasks, a 120. ppm standard solution of analyte and a solution containing an unknown amount of analyte. Ten mL of the unknown solution were added to each flask using a volumetric pipet. Increasing amounts of the 120. ppm standard were also added to the flasks, as illustrated in Table 5. Each flask was then diluted to volume with deionized water and thoroughly mixed. *Note that solution shading (see legend) is not meant to indicate the lack of mixing but just illustrate relative amounts.*

Table 5   *Standard Additions*

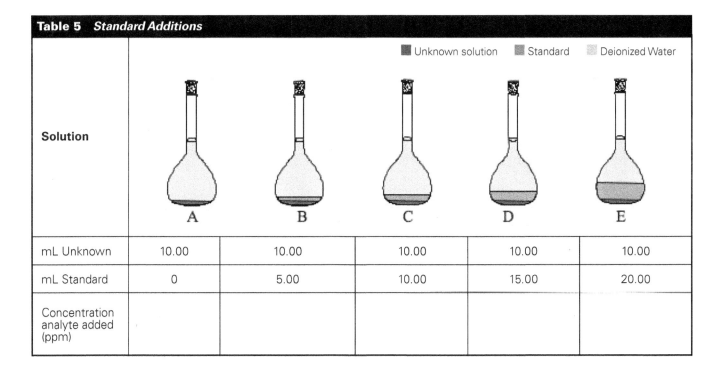

Legend: ■ Unknown solution   ■ Standard   ▨ Deionized Water

Solution	A	B	C	D	E
mL Unknown	10.00	10.00	10.00	10.00	10.00
mL Standard	0	5.00	10.00	15.00	20.00
Concentration analyte added (ppm)					

24. Consider the relative amounts of the three solution components. What happens to the amounts among solutions A-E?

Unknown Solution   Increases   Decreases   Stays the Same

Standard   Increases   Decreases   Stays the Same

Water   Increases   Decreases   Stays the Same

25. As a group, consider the standard addition samples in Table 5. What are the concentrations of analyte *added* to each solution? Fill in the answers in the Table. Check your answers by comparing with another group.

26. How would preparing a 16.0 ppm standard addition sample be different from the standard prepared in Q23?

Consider this…

The atomic absorbance of each solution in Tables 4 and 5 was measured. The resulting graphs are shown below.

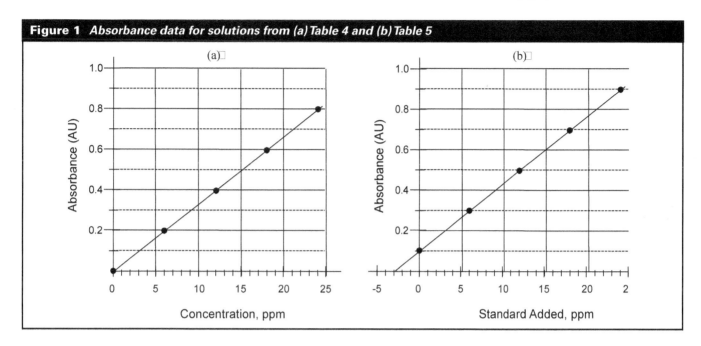

Figure 1 *Absorbance data for solutions from (a) Table 4 and (b) Table 5*

Key Questions

27. Examine Figure 1 carefully. Which graph (a) or (b) corresponds to analysis of the *Standards?* Which corresponds to the *Standard Addition* solutions?

28. Add the absorbances for each solution to Tables 4 and 5, reproduced in part below. Estimate the absorbance values by looking at the graphs in Figure 1.

Table 4 *Calibration Standards*

Solution	1	2	3	4	5
mL Standard	0	8.00	16.00	24.00	32.00
Concentration analyte (ppm)	0	6	12	18	24
Absorbance					

Table 5 *Standard Additions*

Solution	A	B	C	D	E
mL Unknown	10.00	10.00	10.00	10.00	10.00
mL Standard	0	5.00	10.00	15.00	20.00
Concentration analyte added (ppm)	0	6	12	18	24
Absorbance					

29. Based on the absorbances of the five solutions in each graph, which solution has the higher concentration of analyte for the same concentration value on the x-axis?

30. If the absorbance of an unknown solution of the analyte is measured, the concentration of that solution can be determined from the calibration standards. If an unknown solution has an absorbance of 0.50, what is the concentration of analyte in the unknown solution?

31. Consider the 12 ppm solution. The absorbances are different for the calibration standard solution and for the standard addition solution containing 12 ppm analyte.

    a. By how much is the absorbance of the 12 ppm standard addition solution greater than the 12 ppm calibration standard solution?

    b. Examine the absorbances of the other solutions in the calibration standards and standard addition datasets. How much greater is the absorbance for these other solutions?

    c. Where does the "extra" analyte come from?

32. What is the absorbance of the standard addition solution containing no added standard in Figure 1b?

33. Use the calibration graph in Figure 1a to determine the concentration of analyte in a solution with this absorbance.

34. Circle the concentration (x-value) that corresponds to zero absorbance on the standard addition curve (Figure 1b).

35. Compare your answers to Q33 and Q34. What is the mathematical relationship between these two results?

In laboratory situations where standard addition is needed it is unlikely that the slopes of the two plots in figure 1 would have the same value. In this activity the same slopes were used to illustrate the mathematical relationship required to analyze standard addition data.

---

36. Ordinarily, data for either standards *or* standard addition are collected. If you had only standard addition data, explain how to find the *concentration* of an unknown analyte from standard addition data.

37. You are training a new laboratory technician. How do you teach them about when standard addition methods are necessary?

Applications

38. Calculate the concentration for each sample and duplicate in Table 3.

   (a) Is the concentration of any sample below the MDL? If so, what would you report as the result for that sample?

   (b) Is the concentration of any sample above the upper limit of the dynamic range? If so, what should be done for that sample?

39. The EPA method used as the basis of this activity says that the MDL is determined by "processing seven replicate samples through the entire method." These seven replicates are prepared "using reagent water (blank) fortified at a concentration of two to three times the estimated [IDL]." The MDL is calculated, according to the EPA method, by multiplying the standard deviation for the seven replicates by the Student's t value for a confidence level of 99% and for n-1 degrees of freedom. The Harris text specifies that the LOD (=MDL) is found by 3*s/m, where s is the standard deviation of n≥7 replicates and m is the slope of the calibration line. Comment on the difference between these two methods of determining the MDL.

40. For each experimental consideration listed in the Table below, circle the method (Standards or Standard Addition) that is better in your opinion. Give a reason for your choice.

Experiment Consideration			Reason
# unknowns determined per set of solutions	Standards	Standard Addition	
Time required to prepare solutions	Standards	Standard Addition	
Accuracy obtained when a matrix effect exists	Standards	Standard Addition	
Cost of materials	Standards	Standard Addition	
Cost of labor	Standards	Standard Addition	

41. Considering all the factors in Q40, which method (standards or standard addition) is better under what circumstances? *Make a recommendation for your supervisor regarding future analyses done in your lab.*

42. Suppose you wanted to determine the amount of calcium in milk. Which approach is easier: to mimic the milk matrix in your standards or to add standard to milk and use the existing matrix?

43. Consider the shape of most atomic absorption calibration curves (see examples below). Why is it important to use solutions that have concentrations in the working range?

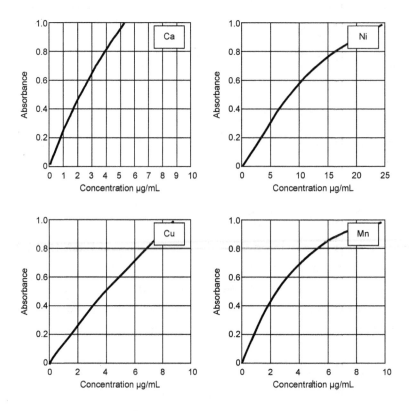

44. Sketch the absorbance vs. concentration graph expected when the standard addition method is used in visible spectroscopy. Explain how to find the concentration of the unknown from this graph.

45. (a) The calibration graph at the right was obtained by AAS using solutions containing known amounts of Zn^{2+}. An unknown sample was measured under the same conditions and the absorbance was 0.560. Determine the Zn^{2+} concentration in the unknown sample.

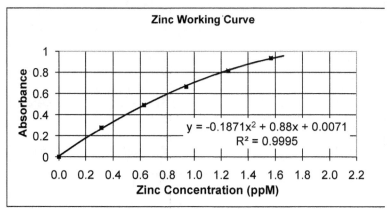

Zinc Working Curve

$y = -0.1871x^2 + 0.88x + 0.0071$
$R^2 = 0.9995$

(b) Explain the difference between the method used to determine Zn in problem (a) and the standard addition method. How would you prepare the Zn solutions if you wanted to use the standard addition method instead? Be specific.

(c) The Zn-containing unknown in problem (a) was diluted prior to analysis in three serial dilutions: 10.00 mL of the original solution was diluted to 100.00 mL; 20.00 mL of the resulting solution was diluted to 100.00 mL and 20.00 mL of that solution diluted to 250.00 mL. What was the Zn concentration in the original solution?

46. A standard addition analysis of a Pb-containing solution was done using AAS. The solutions were prepared by adding 15.00 mL of a diluted unknown Pb solution to a varying amount of 505.8 ppm Pb stock solution, as depicted in the table below. Note that deionized water was NOT added to these solutions to give a constant total volume.

Sample	Diluted Unknown Solution (mL)	505.8 ppm Pb standard (mL)	Absorbance
1	15.00	0.00	0.019
2	15.00	0.09	0.181
3	15.00	0.29	0.440
4	15.00	0.59	0.663
5	15.00	0.81	0.749
6	15.00	1.00	0.798

(a) Plot the absorbance of the solution as a function of ppm Pb added and determine the concentration of Pb in the diluted unknown solution. What is the uncertainty in this Pb concentration?

(b) The unknown solution was prepared by serial dilution as follows: 10.00 mL of the original unknown solution was diluted to 100.00 mL. The resulting solution was diluted serially 10.00 mL to 100.00 mL three more times. What is the concentration of Pb in the original unknown solution?

(c) Based on the uncertainty in the unknown Pb solution (part a) and the tolerances of the volumetric glassware used in diluting the unknown Pb solution from the original concentration, determine the uncertainty in the concentration of the original unknown solution.

47. Lake phosphorous levels were determined spectrophotometrically using a molybdenum assay (EPA Monitoring and Assessment: Phosphorous, http://water.epa.gov/type/rsl/ monitoring/vms56.cfm). The method of standard additions was employed. The samples were prepared in 25-mL volumetric flasks using lake water and standard phosphorous solution, as illustrated in the table.

Sample	Lake Water (mL)	3.00 ppm Phosphorous standard (mL)	Absorbance
1	15.00	0.00	0.109
2	15.00	1.00	0.246
3	15.00	2.00	0.440
4	15.00	3.00	0.645
5	15.00	4.00	0.789
6	15.00	5.00	0.943

Plot the absorbance of the solution as a function of ppm P added and determine the concentration of P in the lake water sample. What is the uncertainty in this P concentration?

48. Zinc in supplements was determined and the following data obtained. Analyze the results and determine whether standard addition methods are required.

Solution Number	Concentration Zn Standard (ppm)	Absorbance
1	0.00	0.002
2	0.30	0.209
3	0.60	0.423
4	0.90	0.601
5	1.20	0.811
6	1.50	0.987
Solution Number	Solution Description	Absorbance
9	Check Standard (0.9 ppm)	0.637
10	Zero standard	0.002
11	Sample	0.548
12	Duplicate	0.572
13	Spike (0.6 ppm added)	0.922
14	Check Standard (0.9 ppm)	0.624
15	Zero standard	0.003

49. The *Standard Addition* solutions A-E were prepared by placing 10.00 mL of the unknown solution into each of five 100-mL volumetric flasks. By what factor has the unknown been diluted when the volumetric flasks are diluted to volume?

Appendix A: Glossary of Definitions

(Excerpt from EPA Method 200.9 Section 3)

Calibration Blank - A volume of reagent water acidified with the same acid matrix as in the calibration standards. The calibration blank is a zero standard and is used to auto-zero the AA instrument (Section 7.10.1).

Calibration Standard (CAL) - A solution prepared from the dilution of stock standard solutions. The CAL solutions are used to calibrate the instrument response with respect to analyte concentration (Section 7.9).

Dissolved Analyte - The concentration of analyte in an aqueous sample that will pass through a 0.45 μm membrane filter assembly prior to sample acidification (Section 11.1).

Field Reagent Blank (FRB) - An aliquot of reagent water or other blank matrix that is placed in a sample container in the laboratory and treated as a sample in all respects, including shipment to the sampling site, exposure to the sampling site conditions, storage, preservation, and all analytical procedures. The purpose of the FRB is to determine if method analytes or other interferences are present in the field environment (Section 8.5).

Instrument Detection Limit (IDL) - The concentration equivalent to the analyte signal which is equal to three times the standard deviation of a series of ten replicate measurements of the calibration blank signal at the same wavelength.

Instrument Performance Check (IPC) Solution - A solution of method analytes, used to evaluate the performance of the instrument system with respect to a defined set of method criteria (Sections 7.11 and 9.3.4).

Laboratory Duplicates (LD1 and LD2) - Two aliquots of the same sample taken in the laboratory and analyzed separately with identical procedures. Analyses of LD1 and LD2 indicates precision associated with laboratory procedures, but not with sample collection, preservation, or storage procedures.

Laboratory Fortified Blank (LFB) - An aliquot of LRB to which known quantities of the method analytes are added in the laboratory. The LFB is analyzed exactly like a sample, and its purpose is to determine whether the methodology is in control and whether the laboratory is capable of making accurate and precise measurements (Sections 7.10.3 and 9.3.2).

Laboratory Fortified Sample Matrix (LFM) - An aliquot of an environmental sample to which known quantities of the method analytes are added in the laboratory. The LFM is analyzed exactly like a sample, and its purpose is to determine whether the sample matrix contributes bias to the analytical results. The background concentrations of the analytes in the sample matrix must be determined in a separate aliquot and the measured values in the LFM corrected for background concentrations (Section 9.4).

Laboratory Reagent Blank (LRB) - An aliquot of reagent water or other blank matrices that are treated exactly as a sample including exposure to all glassware, equipment, solvents, reagents, and internal standards that are used with other samples. The LRB is used to determine if method analytes or other interferences are present in the laboratory environment, reagents, or apparatus (Sections 7.10.2 and 9.3.1).

Linear Dynamic Range (LDR) - The concentration range over which the instrument response to an analyte is linear (Section 9.2.2).

Matrix Modifier - A substance added to the graphite furnace along with the sample in order to minimize the interference effects by selective volatilization of either analyte or matrix components.

Method Detection Limit (MDL) - The minimum concentration of an analyte that can be identified, measured, and reported with 99% confidence that the analyte concentration is greater than zero (Section 9.2.4 and Table 2).

Control Sample (QCS) - A solution of method analytes of known concentrations which is used to fortify an aliquot of LRB or sample matrix. The QCS is obtained from a source external to the laboratory and different from the source of calibration standards. It is used to check either laboratory or instrument performance (Sections 7.12 and 9.2.3).

Solid Sample - For the purpose of this method, a sample taken from material classified as either soil, sediment or sludge.

Standard Addition - The addition of a known amount of analyte to the sample in order to determine the relative response of the detector to an analyte within the sample matrix. The relative response is then used to assess either an operative matrix effect or the sample analyte concentration (Sections 9.5.1 and 11.5).

Stock Standard Solution - A concentrated solution containing one or more method analytes prepared in the laboratory using assayed reference materials or purchased from a reputable commercial source (Section 7.8).

Total Recoverable Analyte - The concentration of analyte determined to be in either a solid sample or an unfiltered aqueous sample following treatment by refluxing with hot dilute mineral acid(s) as specified in the method (Sections 11.2 and 11.3).

Water Sample - For the purpose of this method, a sample taken from one of the following sources: drinking, surface, ground, storm runoff, industrial or domestic wastewater.

Appendix B

Copy of Table 3 to tear off and use for questions 6-20.

Table 3 Solutions Prepared for Atomic Absorption Analysis of Industrial Waste Samples for Lead			
Sample analysis (11.4) with Performance Assessment (9.3 and 9.4)			
Solution Number	Sample Name	Absorbance	Concentration
25	Laboratory Reagent Blank	0.004	
26	Calibration Blank	0.002	
27	Instrument Performance Check (25 ppb)	0.123	
28	Laboratory Fortified Blank (25 ppb)	0.122	
29	Sample 1	0.408	
30	Sample 2	0.521	
31	Sample 3	0.446	
32	Duplicate Sample 3	0.448	
33	Sample 4	0.512	
34	Sample 5	0.926	
35	Sample 6	0.489	
36	Laboratory Fortified Sample Matrix: Sample 6 + 25 ppb Spike	0.568	
37	Sample 7	0.506	
38	Sample 8	0.579	
39	Laboratory Reagent Blank	0.003	
40	Calibration Blank	0.001	
41	Instrument Performance Check (25 ppb)	0.112	
42	Sample 9	0.482	
43	Sample 10	0.009	
44	Sample 11	0.819	
45	Sample 12	0.486	
46	Laboratory Reagent Blank	0.002	
47	Calibration Blank	0.002	
48	Instrument Performance Check (25 ppb)	0.125	

Appendix C

Table to Complete as Quality Assurance Checks are Performed.

Solutions	Criterion for Acceptance	Solutions from Table 1-3 (Q6)	Pass?
Instrument Stability Check	<5% RSD		
Calibration Standards	N/A		N/A
Method Detection Limit	N/A		N/A
Laboratory Reagent Blank	<2.2 x MDL		
Calibration Blank	< MDL		
Samples	N/A		N/A
Replicate Samples	< 10% difference		
Laboratory Fortified Blank	% recovery=85-115%		
Instrument Performance Check	within±5%		
Laboratory Fortified Sample	% recovery=70-130%		

Interlaboratory Comparisons

Learning Objectives

Students should be able to:

Content

- Students will be able to explain how random and systematic error affects experimental results.

- Students will be able to describe what the purpose of an interlaboratory study is and how it might be conducted.

- Students will be able to explain how spike recovery may be used to evaluate accuracy of a method and of a laboratory's performance.

Process

- Students will engage in drawing inferences from data presented in graphical form. (Information processing, Critical thinking)

Prior knowledge

- Students should be able to describe accuracy and precision at least in crude terms.

Further Reading

- Bauer, C.F., S.M. Koza, and T.F. Jenkins 1990. "Collaborative Test Results for a Liquid Chromatographic Method for the Determination of Explosives Residues in Soil" J. Assoc. Off. Anal. Chem. 73: pp.541-552. Accepted as a standard method by Association of Official Analytical Chemists (991.09), ASTM (D5143-90), and EPA (SW-846 Method 8330).

- Harris, D. C. 2007. Quantitative Chemical Analysis, 7th Edition, Chapter 3-5. New York: W.H. Freeman.

- Skoog, D. A., D. M. West, F. J. Holler, and S. R. Crouch. 2004. Fundamentals of Analytical Chemistry, 8th Edition, Chapter 5-7. Belmont, CA: Thompson Brooks/Cole.

Author

Christopher F. Bauer

Consider this...

New analytical methods are usually developed in one particular laboratory by a small number of scientists. If the method is to be useful, the results need to be reliable when used by large numbers of scientists working in many different laboratories. One way of evaluating how well a method performs when applied by many users is called an *interlaboratory study.*

In an interlaboratory study, several different laboratories carry out an analysis of the same materials, and then the results are compared. If all laboratories follow the same procedure, then the comparison provides information about how well the procedure performs under realistic conditions.

The data discussed here are from an actual interlaboratory study of a method for measuring Army munitions residues in soils. The method involves extracting 2 grams of soil with 10 mL of acetonitrile (CH_3CN) for 18 hours followed by filtration and analysis of the filtrate by liquid chromatography. There are several different munitions substances present in munitions waste. Liquid chromatography is used to separate these substances. Ultraviolet absorption spectrometry is used to detect and quantify the amounts of each substance.

Two analytes will be considered: RDX and tetryl.

RDX

Tetryl

Four portions of the same clean soil were spiked with these analytes (i.e. exactly known amounts of analyte were added to the soils in solution form in a solvent that evaporates easily). The concentrations were chosen so that there would be an identical pair of high concentration spikes and an identical pair of low concentration spikes, with about 10% difference in concentration between the two. The laboratories (numbered 1-7) did not know the actual concentrations (mass of analyte per mass of soil). This is called a blind analysis. The reported results are in Tables 1 and 2.

Table 1 *Reported concentrations of RDX (µg/g)*

	Low spike 1	Low spike 2	High spike 1	High spike 2
	90.4	90.4	100.4	100.4

Lab ID	Concentration found by using the method			
1	87.9	85.2	98.3	96.3
2	87.4	88.2	99.7	98.9
3	87.8	83.6	99.6	94.4
4	85.7	87.2	91.6	92.7
5	102.0	102.0	112.0	105.0
6	56.9	63.8	71.7	63.8
7	93.5	91.5	101.0	101.0

Table 2 *Reported concentrations of Tetryl (µg/g)*

	Low spike 1	Low spike 2	High spike 1	High spike 2
	20.1	20.1	25.2	25.2

Lab ID	Concentration found by using the method			
1	17.9	15.0	13.4	8.3
2	19.2	20.0	24.1	23.7
3	16.8	11.3	7.3	12.8
4	18.6	18.8	23.3	24.2
5	4.2	4.1	2.7	3.4
6	14.4	13.6	11.3	18.5
7	17.8	16.7	22.7	22.1

The graphs at the end of this activity—one for each analyte—were created from the data in the tables. You may never have seen information plotted in this way before, so it may take you a while to make sense of the graphs. The measured result for High Spike 1 is plotted against Low Spike 1, and High 2 vs. Low 2. This type of data display is called a *Youden two-sample plot* (one sample result plotted against a second sample result).

Key Questions

1. Look at the RDX graph. Each lab's results are indicated by a label on the points. How many labs contributed points to the graph? Why are there two points for each lab?

2. Instead of plotting in concentration units, the results are plotted as a ratio to the actual known concentration. Use the data in the table to calculate and verify the coordinates of the two Lab 6 data points on the RDX graph.

3. Explain why the numerical values for the axes on the graph have no units.

4. Describe what it means for a data point to fall exactly on the value (1.0, 1.0).

5. Each member of your group should take a turn to restate in different ways what the location of the Lab 6 points tells you about Lab 6. Do this in terms of the position of the two points relative to each other and to the (1.0,1.0) axis point.

6. Agree on and write down a conclusion regarding Lab 6's performance by using the two terms "random" and "systematic" to describe the performance.

7. Look for the Lab 5 data. Describe what you see. Discuss the location of the points in terms of the two terms "random" and "systematic". Agree on and write down a conclusion regarding the performance of lab 5 using the terms "random" and "systematic."

8. Reconsider the meaning of the Lab 6 and Lab 5 data points on the RDX graph and describe their position in terms of the words "percent recovery."

9. Look at the data for all other labs. Describe how each of these labs compares with Labs 5 and 6 in terms of random and systematic error and percent recovery. Note that there is an expanded version of the RDX graph which may help in answering this question.

10. Look up the terms random and systematic in your textbook to determine whether the way you have been using the terms is consistent with formal definitions.

11. What aspect of the location of points tells you about the accuracy of the work of an individual lab? What aspect of the location of points tells you about the precision of the work of an individual lab?

12. Discuss and write a written summary as a group of how a Youden two-sample plot shows evidence about the performance *of individual laboratories*.

Consider this...

The same labs also measured the quantity of tetryl in the same samples using the same method. Remind yourself of the structures of the two analytes. See the two tetryl graphs (full and expanded).

Key Questions

13. Describe the performance of each *individual lab* in terms of its accuracy and precision for measuring tetryl. (Each group member should take responsibility for doing this for two of the laboratories. Then, each group member should read his/her descriptions to the other group members and seek agreement about the description. The group should make a list and turn that in.)

14. Describe the performance of each individual lab in terms of random and systematic error and percent recovery. (Follow same procedure)

15. Write down a list of the procedural steps in the analytical method. If any steps are unfamiliar, refer to your textbook or consult your instructor.

16. Look at and compare the Lab 6 percent recoveries for RDX and for tetryl on the graphs. For both RDX and tetryl, the recoveries are in the range of about 60-70%. This suggests that a *common source of systematic error is affecting every measurement* for both analytes.

Each person in the group should individually try to identify and justify one specific reason—*that lies with the lab*—that could cause these observed recoveries of BOTH analytes. There must be a causal link between your reason and the location of the points on the graph. Then share and discuss your justifications and write down a list.

17. Identify and justify one possible specific reason—that lies with the lab—for why the Lab 5 results for RDX and tetryl are where they are on the graphs.

Consider this...

Consider the possible locations of points on a Youden plot relative to the axes.

A point in Quadrant 1 (upper right) is there because the lab had recoveries that were too high on both samples. A point in Quadrant 3 (lower left) is there because the lab had recoveries that were too low on both samples. Thus, points lying on this diagonal provide evidence regarding the level of systematic error.

A point in Quadrant 2 (upper left) is there because the first result was too low and the second too high. A point in Quadrant 4 (lower right) is there because the first result was too high and the second too low. Thus points lying away from the diagonal provide evidence regarding the level of random error. Note that random error is equally likely to cause results that are both low, or both high.

Turn your attention to the performance of the analytical method, start to finish. Remind yourself of the structures of the two analytes and the steps in the method.

Key Questions

18. Describe the performance of *the method* as a whole in terms of its success in providing reliable results for tetryl. You will need to compare the results *for tetryl* with the results for RDX. Agree on and write down your description, and the supporting evidence.

19. Identify and justify two possible specific reasons—that lie with the method—for the poorer performance on tetryl relative to RDX.

20. Discuss and write a summary as a group of how a Youden two-sample plot shows evidence about the performance *of an analytical method.*

21. As a group, summarize your perceptions. Were all group members improving in their ability to make sense of the graphical data?

Application

22. Assume that 10 labs were involved in an interlaboratory study of a new method. Sketch Youden two-sample plots for each of these cases. Each group member should take one case and then describe his/her graph to the others.

a. small random error and a wide range of systematic errors for labs

b. large random error and small systematic error for every lab

c. small random error and modest systematic error for every lab

d. small random error and large systematic error that affects every lab in a similar way.

23. A Youden two-sample plot requires only two samples per lab. We used two pairs of two samples. What information is retained and what information is lost by analysis of only two samples per lab?

24. As a group, outline and write down the steps needed to carry out an interlaboratory study that leads to constructing a Youden two-sample plot.

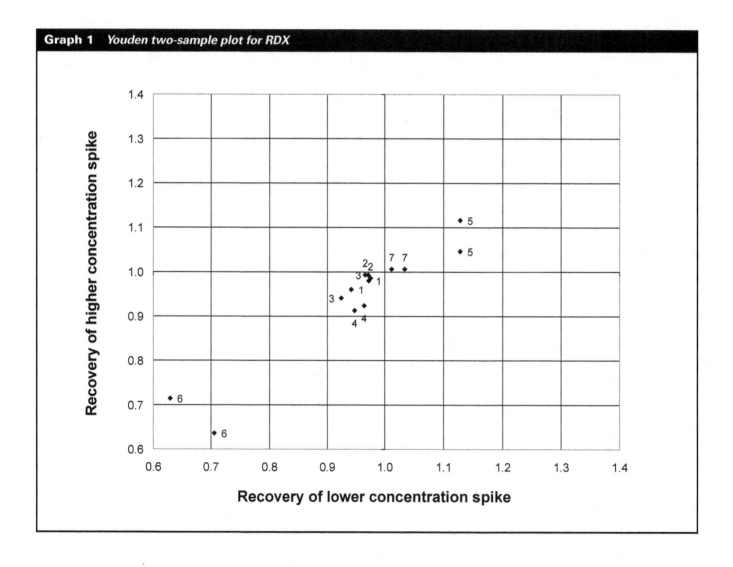

Graph 1 *Youden two-sample plot for RDX*

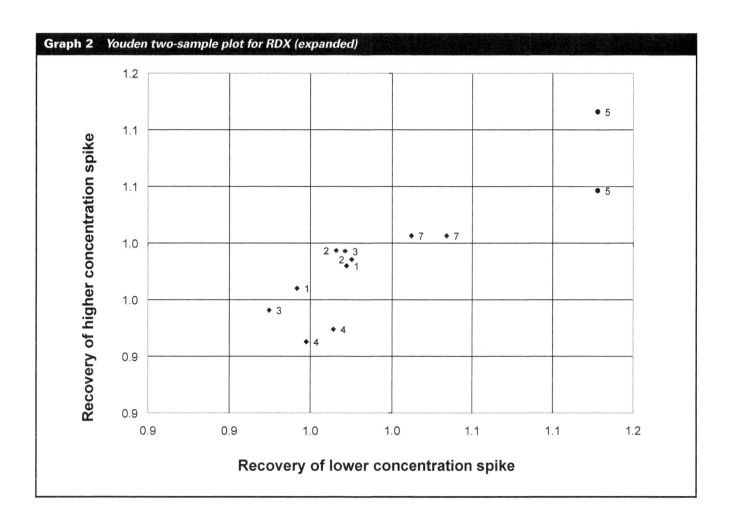

Graph 2 *Youden two-sample plot for RDX (expanded)*

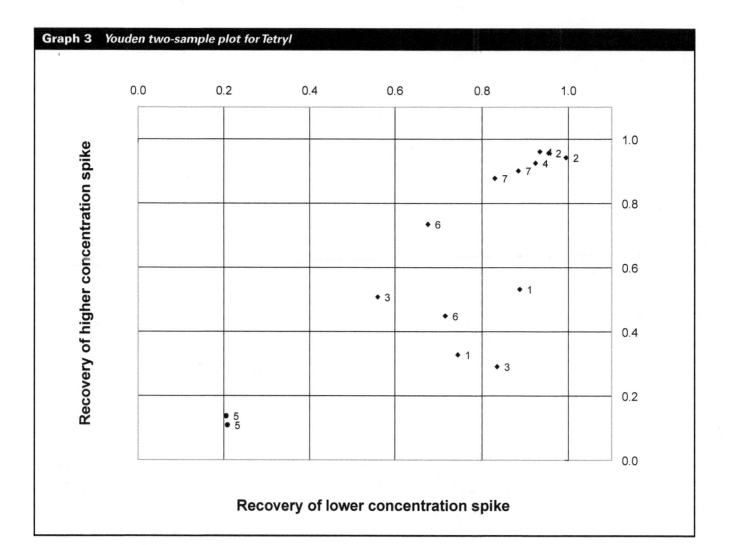

Graph 3 *Youden two-sample plot for Tetryl*

Recovery of higher concentration spike

Recovery of lower concentration spike

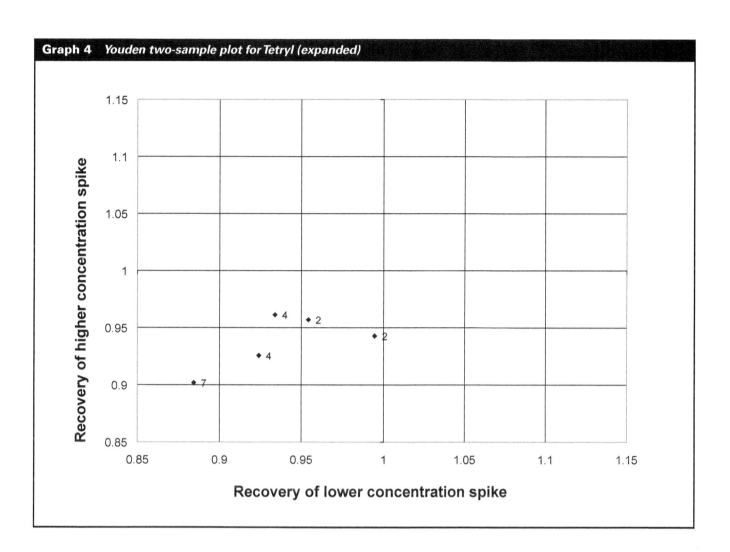

Graph 4 *Youden two-sample plot for Tetryl (expanded)*

Errors in Measurements and Their Effect on Data Sets

Learning Objectives

Students should be able to:

Content

- Classify different sources of experimental error and determine their effect on the mean and dispersion of the data.
- Develop the ability to judge whether an analytical balance has been used correctly.
- Define situations in which it is reasonable to reject data.

Process

- Develop critical skills in assessing the quality of data: Are they "good" or "bad"?
- Identify qualitative differences among data sets.
- Develop generalizations from tabulated data.

Prior knowledge

- Familiarity with simple statistics terms such as mean, range, and standard deviation.
- Familiarity with an analytical balance that can be tared. In particular, students need to know that closing the draft shield improves the reproducibility of the balance, and that hot objects placed on the balance pan create convection currents that reduce the measured value of the mass. They should also know that analytical balances need to be leveled for correct operation.

Further Reading

- Harris, D.C. 2007. *Quantitative Chemical Analysis,* 7th Edition, 42. New York: W.H. Freeman.
- Taylor. John R. 1997. An Introduction to Error Analysis, 2nd Edition, 94. Sausalito CA: University Science Books.

Author

Carl Salter

PGLANA004009a POGIL

Consider this...

As scientists we often must make a large number of repeated measurements. It is usually not practical to communicate or transmit every single measurement to other scientists who are interested in the experiment; instead, we summarize the data set using carefully selected statistics. There are two key aspects of the data: one is the central tendency, or center of the data, which is a value about which the data seem to "cluster," and the other is the spread or dispersion of the data. To summarize the center of the data we usually use the average (or mean) value. The average is the sum of all measurements divided by the number of measurements. To summarize the dispersion of the data we can use either the range or the standard deviation. The range is simply the maximum minus the minimum. The standard deviation is more complicated, and involves the sum of the squares of the differences between each measurement and the average.

Here are formulas for the average, the range, and the standard deviation.

$$\bar{x} = \frac{1}{n}\sum_{i=1}^{n} x_i, \quad r = x_{max} - x_{min}, \quad s = \left(\frac{1}{n-1}\sum (x_i - \bar{x})^2\right)^{1/2}$$

Since 1982, U.S. pennies have been minted from 97% zinc stock electroplated with a thin coat of copper; a typical new penny weighs 2.508 grams. Prior to 1982, pennies were made almost entirely of copper, and weighed more than 3 grams.

Suppose that for several months, a professor has collected lots of pennies from his loose change to use in an experiment that helps students learn to use analytical balances. He removes the pennies that are dated 1981 or earlier and keeps the more recent pennies for the experiment. The pennies are divided into sets of 20.

On the day of the experiment, a trained laboratory assistant takes one of the sets of 20 pennies and weighs them on an analytical balance with a digital display that weighs to 0.0001 grams. Each student then weighs a different set of 20 pennies and records the weights in a laboratory notebook. The students also use analytical balances that weigh to 0.0001 grams.

The students are asked to compare their data with the data obtained by the lab assistant.

Key Questions

1. The following table contains three sets of penny weights; the first set is the set of measurements made by the lab assistant. The other two sets were obtained by students. The data are listed in the order in which the measurements were made. In addition to the raw data, the table contains the summary statistics for each set: average, range, and standard deviation. Compare the three data sets.

	Masses (grams) of pennies		
	Lab asst	**Set 1**	**Set 2**
	2.5283	2.4814	2.5696
	2.4936	2.4782	2.4200
	2.5442	2.4989	2.5234
	2.4892	2.4948	2.4650
	2.5099	2.5065	2.4597
	2.5058	2.4392	2.5110
	2.5580	2.4946	2.5789
	2.5534	2.4932	2.5519
	2.5035	2.5332	2.4505
	2.5058	2.5177	2.5205
	2.5264	2.5423	2.4875
	2.4936	2.4826	2.4896
	2.5038	2.4801	2.5029
	2.5238	2.5022	2.4534
	2.4911	2.4989	2.5559
	2.4892	2.4392	2.5096
	2.4924	2.5470	2.5013
	2.5058	2.5424	2.5673
	2.5580	2.4925	2.4850
	2.5534	2.4910	2.4945
Average	2.5165	2.4978	2.5049
Range	0.0688	0.1078	0.1588
Std deviation	0.0248	0.0293	0.0438
Explanation			

There are important differences between the lab assistant's data and the data obtained by the students. Which of the following explanations fits the differences among the sets? Write your group's choice on the bottom of the table under each of the three columns of data.

Warm object: A student keeps all the pennies in his pocket as he weighs them; as a result, the pennies are above room temperature when they lie on the pan and give low measurements.

Draft: One student doesn't completely close the draft shield as she weighs all 20 pennies.

Wear and tear: Different pennies will receive different levels of wear, dirt, oil, and corrosion while in circulation.

Using a complete sentence, write down a reason for your group's selection for each set.

Lab asst.:

Set 1:

Set 2: draft, we see average, std deviation beans fluuation

Here are three more sets of data obtained by students who made some mistakes. Compare them to the set from the lab assistant, which is listed again for your convenience. Examine the sets and match them to an explanation.

		Masses (grams) of pennies		
	Lab asst	Set 3	Set 4	Set 5
	2.5283	2.5132	2.4969	2.4236
	2.4936	2.5287	2.4864	2.5144
	2.5442	2.4944	2.5453	2.4879
	2.4892	2.5238	2.4812	2.4911
	2.5099	2.5175	2.5044	2.5058
	2.5058	2.4892	2.5005	2.5535
	2.5580	2.5099	2.4746	2.5533
	2.5534	2.5085	2.4864	2.5020
	2.5035	2.5132	2.4978	2.5058
	2.5058	2.5535	2.5455	2.5257
	2.5264	2.4502	2.4958	2.4924
	2.4936	2.5020	2.5207	2.4892
	2.5038	2.9464	2.4069	2.5124
	2.5238	2.5257	2.4955	2.5257
	2.4911	2.5175	2.5095	2.4936
	2.4892	2.5264	2.4812	2.4911
	2.4924	2.4879	2.5019	2.5132
	2.5058	2.4780	2.5005	2.5099
	2.5580	2.4936	2.4069	2.4502
	2.5534	2.5085	2.4969	2.5056
Average	2.5165	2.5294	2.4917	2.5023
Range	0.0688	0.4962	0.1386	0.1299
Std deviation	0.0248	0.1006	0.0344	0.0295
Explanation				

Not level: One of the balances is not level; as a result it gives measurements that are slightly low.

Not tared: One student fails to tare the balance when she weighs her first penny, but thereafter she tares the balance between each weighing.

Typo: A student transposes the second and third digits of one value as he records it in his notebook; that is, he records 2.94xx instead of 2.49xx.

Using a complete sentence, write down a reason for your group's selection for each set.

Set 3: Not level : becaus 2.9464 too high.

Set 4: Not level : the averay is low.

Set 5: Not tared

2. Which error had the greatest effect on the average value? How do you know? Why did it have a big effect?

Set 4, becaus auery value low

3. Are the range and standard deviation consistent? That is, for the six sets of data, do the range and standard deviation show similar trends regarding the dispersion of the data? Explain.

4. Which error had the greatest effect on the dispersion of the data? How do you know? Why did it have a big effect?

5. Compared to the lab assistant's data set, what happens to the statistics of the other sets of data? Your group should discuss each data set. In the table below, summarize the changes in the average and the standard deviation for each cause. Indicate whether the statistic was affected, and whether it increased or decreased. Use up or down arrows to indicate increase or decrease.

Explanation/cause	Average	Std Deviation
Warm pennies weighed		
Draft shield open		
Not level		
Not tared for first penny		
Typo: transposed digits for one value		

Consider this…

All measurements are subject to error. Different sources of error will affect the data set differently. An experimental error will usually have one of the following effects:

1) Affect just one measurement in a unique way, which to some degree can change both the average and the dispersion,

   or

2) cause an increase in ~~(distribution)~~ dispersion of the data by affecting all the data in unpredictable, random ways,

   or

3) cause all the data to shift either to higher or lower values by about the same amount, which, of course, will have an effect on the average. This effect can be eliminated if the source of the error can be found.

some error came dispersion change

Key Questions

Q6 - Q8 refer to the information about effects of error given above.

6. Will the second type of error affect the mean? If so, how will it affect the mean?

No, it only affect distribution, not average (or mean)
mean

7. How will the third type of error affect the dispersion of the data as measured by the standard deviation?

average (mean)
distribution
average
5 *if*
If the shift change, if affect the average, became it change location.

8. If an instrument is used properly, what type of error will still be present in the data?

Now apply the information about effects of errors to Sets 1-5 of the penny data.

9. Assign causes to an effect.

Effect	Explanation/cause (from table above)
Affect average and dispersion by changing one datum	
Increase dispersion by altering all data	
Shift all values, changing the mean	

10. Which of these three types of error would you call a blunder? Which is a systematic error or bias? Which is random error?

like mistake.

11. In a complete sentence, summarize the effect of a blunder on the summary statistics of the data.

12. Summarize the effect of a systematic error on the summary statistics of the data.

13. Summarize the effect of random error on the summary statistics of the data.

14. Why is random error always present even when an instrument is used correctly?

15. Explain how you could spot a systematic error or a blunder. Which of these two types of error is easier to spot?

16. What types of error require a student to discard data and make a new set of measurements?

17. Many students will say that a data set has been affected by "human error," but most scientists do not accept "human error" as a type of experimental error. What do you think "human error" means, and what effect would it have on a data set?

As a group, discuss the three types of errors and decide which, if any, are "human error."

Applications

18. a. On many rulers, the zero mark for the measurement scale is not at the physical edge of the ruler. What kind of error will result if someone uses the ruler and doesn't notice this? What will be the effect on the data?

b. When you use a ruler, you sometimes have to estimate the length between scale markings. If you do this several times, your estimate will probably not be the same every time. What kind of error is this? Explain its effect on the data.

19. Very often, scientists will check or calibrate an instrument to make sure it is making correct measurements on an accepted standard sample. What type of error does this practice reduce or eliminate?

20. Modern instruments usually have digital readouts; the reading in the rightmost digit will sometimes fluctuate up and down among several values. What type of error is this? Explain its effect on a large data set.

21. What type(s) of error(s) does the control experiment reveal? Explain.

22. Suppose that you can reproduce a short-time physical or chemical event over and over again. Discuss possible sources of errors associated with measuring the duration of the event using a stopwatch. ("Human error" is not a source.) What effect do the sources of error have on the data?

23. a. How do scientists measure random error? What statistical values express the amount of random error in a data set?

b. How do scientists measure systematic error? What statistical values express the amount of systematic error in a data set?

24. What are the advantages and disadvantages of using the range to estimate dispersion? Contrast it to the standard deviation as a measure of dispersion.

25. Suppose you are a teaching assistant for the penny experiment, and your job is to review the student's data for possible problems. What would you expect to see in a student's data set for the following problems? What course of action would you suggest to the student?

a. A student misreads the balance and puts the decimal point in the wrong place for all 20 measurements.

b. One student weighs all 20 pennies without checking to make sure the balance reads zero between each weighing.

c. One student misunderstands the instructions and weighs one penny 20 times.

d. A student doesn't want to wait for an analytical balance, so he finds a balance that weighs to 0.01 grams and uses that balance to weigh the pennies.

e. A student gets one penny that was minted early in 1982, when some pennies were still made with 95% copper.

26. "Your data are precise but inaccurate." What have you just been told about the random error and systematic error in your data?

27. "Your data are imprecise but accurate." What have you just been told about the random error and systematic error in your data?

28. Suppose you and two lab partners need to weigh six pennies quickly; you have access to three balances so each of you weighs two pennies. Suppose exactly one of the six pennies is an old one made of copper, but no one in your group knows about the change in penny composition that took place back in 1982. Would the weight of this penny look like a blunder, a systematic error, or a random error? How would you try to determine the source of the error? Ultimately, how would you convince yourself that one penny really weighs more than the others?

The Gaussian Distribution

Learning Objectives

Students should be able to:

Content

- Explain how the shape and position of a Gaussian distribution are controlled by the mean and standard deviation.

- Explain how histograms can be used to illustrate the frequency and distribution of data, and how the sample size affects the accuracy of the histogram.

- Describe the relationship between the Gaussian distribution function and probability.

Process

- Interpret graphs of data, especially histograms. (Information processing)

- Estimate the distribution parameters of a data set. (Information processing)

- Draw conclusions about data sets of different distributions and sample sizes. (Critical thinking)

Prior knowledge

- Familiarity with common statistics formulae such as mean and standard deviation.

Further Reading

- Harris, D.C. 2007. *Quantitative Chemical Analysis,* 7th Edition, pp.59-65. New York: W.H. Freeman.

- Taylor, J.R. 1997. *An Introduction to Error Analysis,* 2nd Edition, p.121. Sausolito, CA: University Science Books.

- Bevington, P.R. and D.K. Robinson. 1992. *Data Reduction and Error Analysis for the Physical Sciences,* 2nd Edition, 28. Boston: McGraw-Hill.

- "Gaussian function", http://en.wikipedia.org/wiki/Gaussian_function, accessed June 29, 2011.

Author

Carl Salter

Consider this...

A statistics class surveyed 218 motorized vehicles in the parking lot of a mall, counting the number of vehicles that had certain characteristics. A particular characteristic defines a class of vehicles; the number of vehicles in that class is the **count** or **frequency** of that class. The data are displayed using histograms: the classes are listed on the horizontal axis, and the frequency of the class is represented by the height of the bar. One advantage of histogram graphs is that it is easy to spot the **mode,** the value that occurs most often. The mode is the class with the highest frequency.

Here are some of the results of the survey based on license plate numbers, passenger seats, and dents. Notice that all three histograms have the same vertical axis, number of vehicles, abbreviated "# of v's."

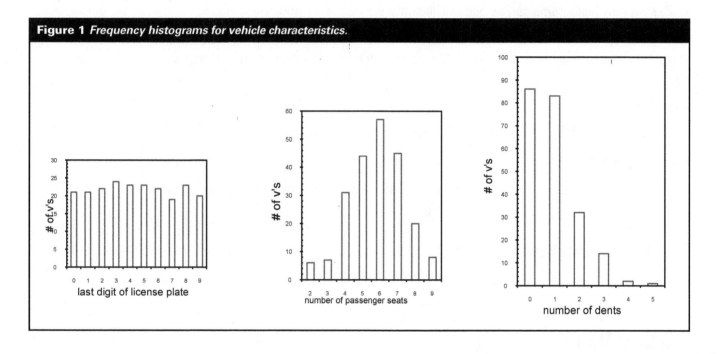

Figure 1 *Frequency histograms for vehicle characteristics.*

Key Questions

1. How many license plates ended with a "0" (zero)?

2. What was the mode of the number of passenger seats among the 218 vehicles?

---

3. Was the number of vehicles without any dents greater than or less than the number with dents? Explain.

4. Assume that the 218 values of the number of passenger seats are in a list sorted from lowest to highest. What would be the first 20 values in the list?

5. Among all three histograms, which class has the greatest frequency?

6. Which histogram has the most classes?

7. Discuss as a group: What result would you get if you added up the heights of the bars in each histogram? Record your best answer here.

There are many ways that random events can be distributed, but the three most common are Uniform, Gaussian, and Poisson. When all outcomes are equally probable, we say that they are uniformly distributed— heads and tails for a coin flip are *uniformly distributed,* as are the results for rolling a six-sided die. Events that appear to be clustered around a single mode are usually assumed to be *Gaussian distributed.* (The Gaussian distribution is also called the *Normal distribution.*) Events that are relatively rare, so that zero occurrences are quite likely, often follow a *Poisson distribution.*

8. As a group, examine the preceding histograms and decide which type of distribution function, Uniform, Gaussian, or Poisson, each follows. Add these labels to the histograms on the previous page.

Generating Probability Histograms

An important transformation that is often performed on histograms is to divide the count or frequency of each class by the total count. This transformation converts the y axis of the histogram to "probability."

Figure 2 contains the histogram for passenger seats converted to probability.

9. Each group member should pick a different value on the original frequency histogram and confirm that it was divided by 218 to give the corresponding result on the probability histogram in Figure 2.

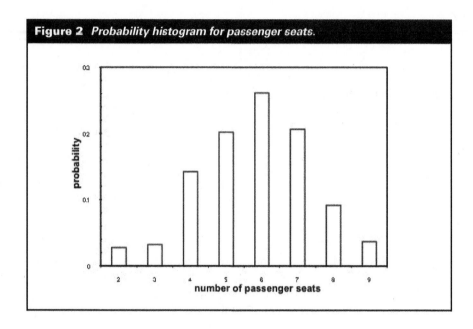

Figure 2 *Probability histogram for passenger seats.*

10. Discuss as a group: What result would you get if you added up the heights of the bars in the probability histogram? Record your best answer here.

The Gaussian Distribution Function

The Gaussian function is the familiar "bell-shaped" curve that peaks at some central value of x and then falls off rapidly to zero as x goes to either plus or minus infinity.

The formula of the **Gaussian distribution**

$$G(x) = \frac{1}{\sigma\sqrt{2\pi}}\, e^{\frac{-(x-\mu)^2}{2\sigma^2}}$$

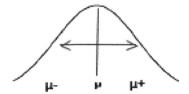

where μ (mu) and σ (sigma) control its position and width, respectively. The Gaussian distribution is a **two-parameter** function; the parameters are μ, **the mean** and σ, the **standard deviation.**

Notice that σ appears in the formula twice.

μ is the value of x where the maximum occurs. σ controls the width of the curve; that is, how rapidly it falls to zero. The points of inflection of the curve occur at $x = \mu \pm \sigma$.

Since the Gaussian distribution function is used to represent probability, the total area under the curve must equal one; that is, its area is **normalized.**

This requirement restricts the value of the height (or amplitude) of the curve:

the height of the curve at the maximum $\dfrac{1}{\sigma\sqrt{2\pi}}$

---

11. Mark the points of inflection on the Gaussian in the previous box.

12. What type of proportionality relationship exists between the height of the Gaussian and its width?

13. As a group, complete the following statement about normalized Gaussians:

A _____ Gaussian is short, a skinny Gaussian is _____.

Consider this...

Here is a plot of three normalized Gaussian distributions. The legend contains the mean and standard deviation of each curve:

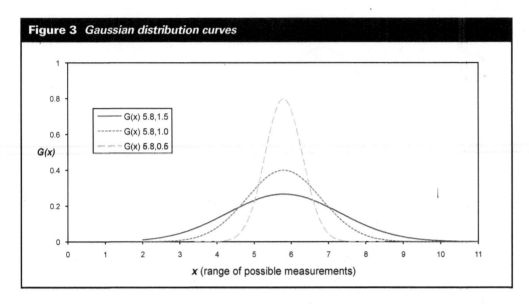

Figure 3 *Gaussian distribution curves*

14. In the legend of Figure 3, which parameter is listed first, μ or σ? Justify your assignment.

15. What is the ratio of the heights of the tallest and shortest Gaussians in Figure 3? What is the ratio of their σs?

16. If the three Gaussians in Figure 3 represent the distributions of three different sets of data, which data set has the **best precision**? Explain your answer.

Linking the Histogram to the Gaussian

The usual formulae for the mean and standard deviation can be used **to estimate** the curve parameters needed to plot a Gaussian distribution function that fits a histogram:

$$\text{mean} \quad \mu = \frac{1}{n} \sum_{i=1}^{n} x_i$$

$$\text{standard deviation} \quad \sigma = \sqrt{\frac{\sum_{i=1}^{n} (x_i - \mu)^2}{n - 1}}$$

17. What is the value of n for the data in the passenger seat histogram?

18. What set of values from the passenger seat histogram would be used as the x_is in the formulae for mean and standard deviation? (Review your answer to Q4.)

19. Suppose you have a histogram where the measurements are weights, so that the x values have units of g. What would be the units of μ, σ, and $G(x)$? Record your group's best answers here.

Consider this...

Figure 4 shows the Gaussian distribution function derived for the passenger seats histogram. The mean and standard deviation of the Gaussian were estimated from the formulae above, rounded to the nearest tenth.

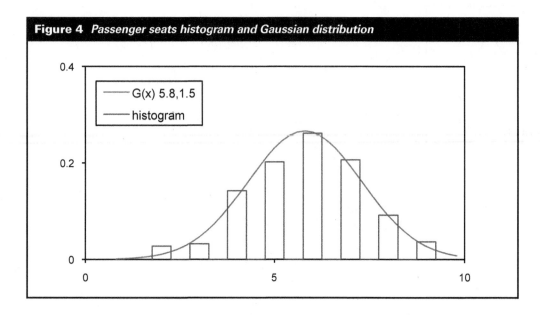

Figure 4 *Passenger seats histogram and Gaussian distribution*

20. Is Figure 4 a frequency or probability histogram? Why is that important if the Gaussian distribution is going to fit the histogram?

21. Here's a way to tell if the Gaussian curve is a good fit for the histogram: The curve should go through the top at least half the histogram bars. How many times does that happen in Figure 4?

22. Debate the following statement: "The Gaussian distribution in Figure 4 is a reasonable approximation of the shape of the histogram." Assess the agreement for both the mean and the standard deviation as well. Record your group's conclusions here.

23. On Figure 4, sketch a Gaussian distribution with a mean that fits the histogram, but with a standard deviation that doesn't, and another Gaussian with a standard deviation that fits, but with a mean that doesn't.

24. As a group, summarize what insights you have gained about the link between a histogram and the Gaussian distribution.

25. Support, refute, or refine the following statement: "A Gaussian distribution is a representation of a set of measurements."

The Number of Measurements and the Uncertainty in the Gaussian Distribution

A measuring system can be idealized as a very large (infinite!) population of values that has a particular mean and standard deviation. Every time we make a measurement we "withdraw" a value from the population. The more times we sample the population, the better we get to know its mean and its standard deviation.

Consider this...

The histograms displayed in Figure 5 were generated by sampling the population defined by the Gaussian distribution with $\mu=5.8$ and $\sigma=1.5$, the same Gaussian we fitted to the passenger seat data. (This Gaussian is plotted on each graph.) But the number of times the distribution was sampled increases as you go down the figure, from just six times for the top histograms to 125 times for the histograms at the bottom. Examine the effect of the sample size n on the appearance of the histograms.

26. As a group, examine the three $n=6$ histograms in Figure 5. Do they appear to be Gaussian-distributed? Record your group's decision here.

27. As a group, examine the other sets of histograms. What is *the lowest value of n* that your group believes reveals the Gaussian distribution of the random numbers?

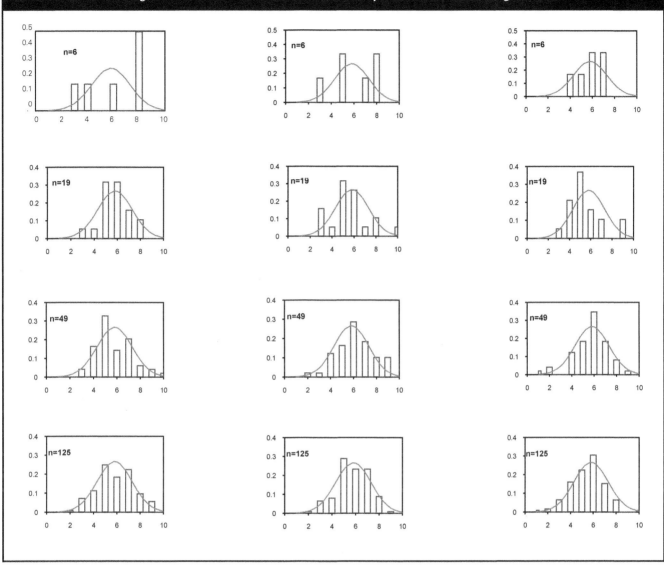

Figure 5 *Probability histograms of Gaussian-distributed random numbers generated using Excel's random number routine in the Data Analysis Toolpak. For all histograms the generating Gaussian distribution was μ=5.8, σ=1.5; this curve is overlaid on each histogram. The number of random numbers sampled to create each histogram is n.*

28. Divide up the work for the following exercise among the members of the group. For each histogram, count the number of times the Gaussian curve passes through the top of a histogram bar. As a group, decide which value of n produces the best fit to the original Gaussian distribution. Record your answer here.

29. If a group of histograms has a similar appearance, then the sampling process that produces them is reproducible. Discuss as a group: Which value of n produces the *most similar* set of histograms?

30. Examine the overall *width* of each histogram and how well it matches the width of the Gaussian. How reliable is an estimate of the standard deviation based on just six measurements? What about other values of n? Record your group's best answer here.

31. How is it possible that the same Gaussian distribution can produce histograms with different shapes? What happens to the shape of the histograms as n increases? Record your group's answers here.

32. Analytical chemists often assume that their data are Gaussian distributed, but they seldom attempt to prove it. Why? Record your group's best answer here.

Estimating Probability using the Area under the Gaussian Distribution

Consider this...

The advantage of deriving a Gaussian distribution function for a set of measurements is that the function can be used *to make predictions about the probability of subsequent measurements,* because the area under the normalized Gaussian is probability.

33. a. What is the probability of getting a measurement with a value from $-\infty$ to $+\infty$?

b. If the normalized Gaussian is integrated from $-\infty$ to $+\infty$, what is the numerical result?

34. a. What is the probability of getting a measurement with a value from μ to $+\infty$?

b. Based on the symmetry of the normalized Gaussian, what is the numerical result if the Gaussian is integrated from μ to $+\infty$?

Because mathematicians and statisticians often need to know the area under sections of the normalized Gaussian, they have set up a table to find the area quickly. (They have a special name for the area under the Gaussian: it is called the "error function," and is abbreviated erf (x).)

To set up a table that can be used for any Gaussian, the measurement coordinate x is expressed in units of the standard deviation. The tabulated values of area are based on the distance to from a particular x to μ. The new coordinate is usually given the symbol z. Its formula is

$$z = \frac{x - \mu}{\sigma}$$

So z is the distance of some value of x from the mean; dividing by σ converts the distance to "units" of standard deviation. Notice that z is dimensionless.

Because the Gaussian distribution is symmetric, the table for the error function only covers positive values of z.

Here is a brief table of areas under the Gaussian based on z. To make the table easier to understand, the normalized Gaussian is graphed as a function of z.

z	erf(z) Area, 0 to z
0	0.0000
0.5	0.1915
1.0	0.3413
1.5	0.4332
2.0	0.4773
2.5	0.4938
3.0	0.4990
∞	0.5000

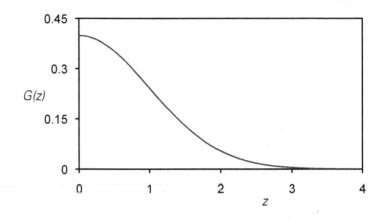

35. Is the final entry in the table consistent with your answer to Q34?

36. Draw vertical lines from the horizontal axis up to the Gaussian at $z=1$, $z=2$, and $z=3$. Compare the areas under the curve from 0-1, 1-2, and 2-3. Order these areas from greatest to smallest.

37. Discuss as a group: Why does the area increase a lot from 0.5 to 1.0, but very little from 2.0 to 2.5? Record your best explanation here.

38. Shade the area under the curve from $z=1$ to $z=2$ on the graph, and compute its area using values from the table.

39. For any Gaussian distribution function, how much probability is there from μ, the mean, to $\mu+1\sigma$, one standard deviation greater than the mean (that is, the right-hand point of inflection)? Using a symmetry argument, determine the probability from $\mu-1\sigma$ to $\mu+1\sigma$ (that is, between the two inflection points).

40. True or false: "There is a 95.5% probability of getting a measurement from $\mu-2\sigma$ to $\mu+2\sigma$." As a group, justify your answer and record it here.

41. Suppose your group measured something 218 times, say the pH of water flowing in a stream. How would you examine your data to determine whether or not it was Gaussian distributed? Record your best answer here.

42. Suppose your group measured something 218 times, say the weight of newly minted pennies. How would you use your data to predict the probability of future measurements on new pennies? Record your best answer here.

43. Debate the following statement: "Without a histogram of the current data, it is impossible to make predictions about the probability of future measurements."

44. As a group, summarize what insights you have gained about the relationship between laboratory measurements and Gaussian distribution functions. Why do scientists find Gaussian distributions useful?

Applications

More about Gaussian Curves

45. In Figure 3, estimate the area under the first and third Gaussians by approximating both curves as triangles. Draw a "best-fit" triangle over each Gaussian. Some parts will not be covered by the triangle; that's okay. (Using the height and the base, the area of a triangle is ½ *hb*.)

Questions 46-49 refer to the following graph:

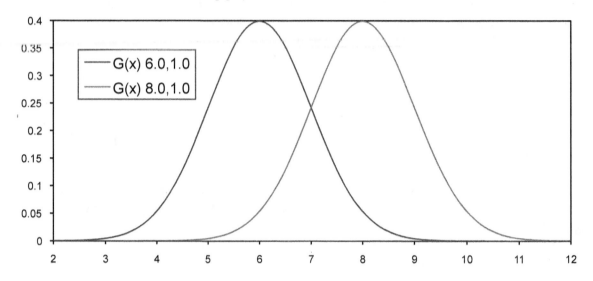

46. What is special about the point of intersection of these Gaussians?

47. Which is greater, the width of one of the Gaussians midway between the top and the bottom (at half its height), or the width of that Gaussian at its points of inflection?

48. Estimate the height of the Gaussians and compare the height to $(2\pi)^{-1/2}$.

49. Describe how you could use your knowledge about the shape of a Gaussian to make a rough estimate of the parameters μ and σ for a Gaussian distribution function that would fit a given histogram.

More about Histograms and Gaussian Distributions

50. For the following vehicle characteristics, discuss the shape of the histogram and the distribution function it would likely follow: Number of wheels and tires, number of windows (individual plates of glass), number of cup holders, number of magnetic decals, and number of steering wheels.

51. Use the following formulae, which involve sums over the classes, to show that the mean and standard deviation of the Gaussian distribution for passenger seats are 5.81 and 1.54 respectively. Estimate the probability of each class from the histogram in Figure 2. The probabilities p_i from the histogram can be used in the formulae:

$$\mu = \sum p_i x_i \quad {}^*\sigma = \sqrt{\sum p_i (x_i - \mu)^2}$$

derived from the usual formulae for mean and standard deviation
* for large n, assume that $n\text{-}1 \approx n$

Histograms of Continuous Quantities

In the histograms for vehicles, the characteristics that were observed were discrete variables that could be expressed as integers. Histograms can be used for characteristics that are continuous; however, each class is defined by a range of continuous values. A decision must be made about how to clump, or "bin", the data for counting. (The classes of a continuous histogram are often called bins.) Consider the following histogram for titration data:

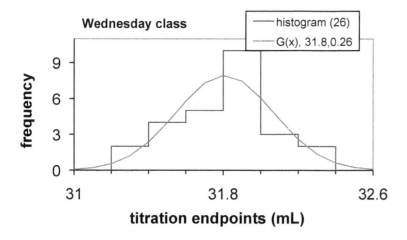

(Notice that this histogram doesn't have bars. This is called a "step" histogram.)

52. How many bins does this histogram have, and what is the width of the bins?

53. What are the units of the standard deviation?

54. How many titrations did the Wednesday class do? How many endpoints were measured between 31.8 and 32.0 mLs?

55. Generate a set of data from 31.0 to 31.8 mLs consistent with the left half of the histogram.

56. What would the histogram look like if only two bins with ranges 31-32 and 32-33 had been used?

57. Compute the mean and standard deviation of the titration endpoints using the formulae in Question 51. By convention, the x value of the bin is its midpoint; for example, the measurements counted in the 31.8–32.0 bin are all assigned the value 31.9 mL.

Some judgment is required to create a histogram for continuous data. If you use too few bins, then you won't see how the data is clustering; if you use too many bins, then the data are separated so much that no structure is revealed. A rough guideline is to let the number of bins equal the square root of the total count. But even that suggestion may have to take a backseat to other considerations. For example, six bins were used for the Wednesday class's titration data so that it could be compared with Tuesday's data:

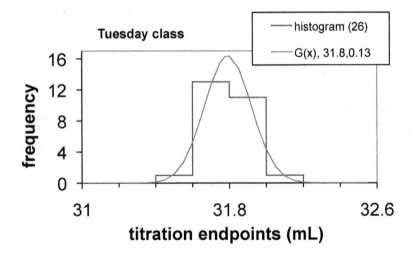

By using the same bin width for both sets, the histograms readily show that the data of Tuesday's class data are more reproducible than that of Wednesday's class.

58. Using information from the histograms, justify the claim that Tuesday's titration results are more reproducible.

59. Based on the data available in the histogram, what is the probability of the Tuesday class getting a measurement in the range 31.8 ±0.13 mLs? Compare that with the probability of the Wednesday class getting a measurement in the same range.

Plotting Gaussian Distribution Functions in Excel

NORMDIST

It is easy to plot normalized Gaussians in Excel because the function is built into Excel. The NORMDIST function (short for NORMal DISTribution) will compute $G(x)$. The format of the function is NORMDIST (x, mean, standard deviation, cumulative), so from Excel's point of view, NORMDIST is a function with four arguments. The first argument is x, the position on the horizontal axis where you want to compute $G(x)$. The x values will usually be stored in a column of cells in the spreadsheet. Generally speaking, there should be about 30 equally spaced values of x, and they should cover the range of the mean $\pm$ 2 standard deviations. The next two arguments are the mean and standard deviation of the Gaussian; that is, the curve parameters. They can be entered as numbers directly into the function, or they can be referenced from cells—this allows them to be changed. The final NORMDIST argument, cumulative, is a "logical" argument. Think of it as a switch: if it equals zero (or false), you get the value of the Gaussian, but if it equals one (or true), you get the area under the Gaussian from zero to x—that is, cumulative =1 actually gives you the erf function.

Directions

Range of x: In cells A1 and A2 enter "0" and "0.25" then select the two cells, and drag down the A column to create a series of values from 0 to 20 with an increment of 0.25. You'll have to drag down to cell A81.

NORMDIST: In cell B1 enter "= normdist (A1,10,1,0)". (A1 can be entered by clicking on the A1 cell if you wish.) When you hit enter, the value 7.7E-23 should appear in B1. Copy this formula down column B to B81. Inspect the numbers that were computed; the maximum value should be 0.399 and it should appear in B41, right next to the x value 10, which is the mean.

Graph: Select cells A1 thru B81. Select the Chart Wizard and select X-Y scatter plot. Select smooth curves without points (data markers). Finish the chart. You should see a narrow Gaussian curve that peaks at $x=10$.

Changing the curve parameters: In cell B1 type "=normdist(A1,12,2,0)". When you hit enter, the value 3.0E-09 should appear. Copy this formula down column B. Inspect the numbers in column B; the maximum should be 0.199 in B49, next to $x=12$. The graph should now appear to be a broad Gaussian with a peak at $x=12$. To change the mean and standard deviation more easily, we should reference them to cells. In cells E1 and E2 enter "mean" and "std dev." In cells F1 and F2 enter "10" and "1". Click on cell B1, and edit its formula—select "12" (the mean) in the formula and then click on cell F1, which makes F1 replace "12", then immediately press the F4 key so that dollar signs appear: F1. Repeat for "2", the standard deviation, but this time click on cell F2. The dollar signs indicate "absolute" cell references, meaning that these cell addresses will NOT change when the formula is copied. The formula in B1 should be now be "= normdist (A1,F1,F2,0)". Copy this formula down column B. You should now see the original curve in the graph. You can now enter different values of the mean and standard deviation in cells F1 and F2 and the curve will immediately change.

Explore: See how small and large you can make the mean and standard deviation and still get reasonable curves. You may also wish to explore the cumulative = true argument of NORMDIST.

60. Here is a probability histogram created from 500 random numbers. Plot the histogram data in Excel, compute the parameters of the Gaussian curve that should fit the histogram, and then use NORMDIST to draw the Gaussian distribution curve on the same graph. The mean and the standard deviation could be estimated by visual inspection, or they could be computed using the equations in Q51. *Because the bin width is not 1.0, the histogram is not yet normalized;* that is, the area under the "step" curve is not 1.0! (The histogram probability is "per bin," while the Gaussian curve's probability is "per unit of *x*") You have two choices to get the curve to fit the histogram: You can either complete the normalization of the histogram by **dividing** each probability by the bin width, or you can **multiply** NORMDIST by the bin width.

Bin midpoint	Probability
51.8	0.002
52.2	0.010
52.6	0.030
53.0	0.014
53.4	0.056
53.8	0.070
54.2	0.102
54.6	0.128
55.0	0.118
55.4	0.130
55.8	0.104
56.2	0.092
56.6	0.064
57.0	0.034
57.4	0.026
57.8	0.008
58.2	0.008
58.6	0.004

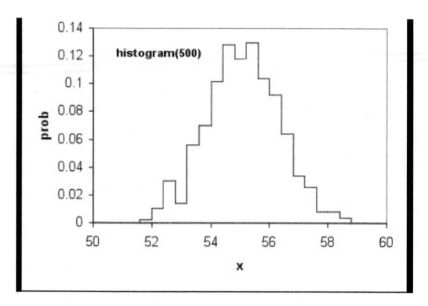

61. Gaussian curves are sometimes used in spectroscopy to approximate the shape of absorption or emission bands. Here the pink absorption band of phenolphthalein has been plotted against wavelength. The absorbance values were recorded at 10 nm intervals from 415 to 635 nm. You could use the equations in Q51 to find the curve parameters; however, it is probably easier to estimate them by visual inspection. The value of the mean is just the top of the curve (spectroscopists call it λ_{max}), and the standard deviation can be estimated by the width from λ_{max} to an inflection point. The challenge here is that the area under the band does not equal 1.0. Since it wouldn't be desirable to normalize the absorption band by dividing by its area, it will be necessary to multiply NORMDIST by the area. The area under the band can be estimated by summing the absorbances and multiplying by the 10 nm interval; this leads to an estimate of about 63. The value for the area, like the mean and standard deviation of NORMDIST, should be referenced to a cell so that it can be adjusted slightly for a best fit. The absorption band is not symmetric, so it clearly will not fit a Gaussian curve perfectly, but with a bit of trial and error you will find that a reasonable compromise fit is possible. You can also try to get just the left side or right side to fit to a tighter Gaussian. You will discover that one side of the band is more Gaussian than the other!

λ (nm)	Abs
415	0.007
425	0.007
435	0.018
445	0.031
455	0.052
465	0.075
475	0.115
485	0.163
495	0.240
505	0.319
515	0.425
525	0.552
535	0.705
545	0.909
555	1.008
565	0.812
575	0.463
585	0.219
595	0.095
605	0.050
615	0.027
625	0.028
635	0.020

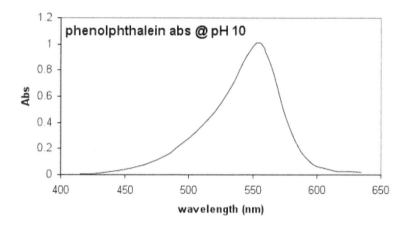

Statistical Tests of Data: The *t* Test

Learning Objectives

Students should be able to:

Content

- Explain how the *t* value changes due to changes in the means, the standard deviations, and the number of data.

- Explain how the *t* test is used to make decisions regarding significance, and be able to use Excel's *t* test output to make such decisions.

- Develop an understanding of the *t* value's relationship to the probability of a significant difference.

Process

- Interpret tables of data. (Information processing)

- Draw conclusions about data sets. (Critical thinking)

- Express concepts in grammatically correct sentences. (Communication)

Prior knowledge

- Familiarity with common statistics terms such as mean, range, and standard deviation.

- Familiarity with simple cases of error propagation, especially sums and differences.

Further Reading

- Harris, D.C., *Quantitative Chemical Analysis*, 7th Edition, 2007 WH Freeman: USA, Harris, D.C. 2007. *Quantitative Chemical Analysis*, 7th Edition, pp.59-65. New York: W.H. Freeman.

- Hecht, H.G. 1990. Mathematics in Chemistry: *An Introduction to Modern Methods*, pp.210-255. Englewood Cliffs, NJ: Prentice Hall.

Authors

Carl Salter

PGLANA004011a

POGIL

Scientists often compare sets of similar measurements. For example, an analytical chemist might compare two different methods for measuring the amount of carbon dioxide in air. Or a food chemist might measure dissolved carbon dioxide in two different brands of soda. Even when the comparison isn't explicit, analysts are often implicitly comparing their measurements against reference samples or accepted standards. Whatever the case, experimental scientists often must decide whether an observed **difference** is important and worthy of communication. Usually they will focus on the two most important statistics about a data set: its average $\overline{X}$ (or mean) and its standard deviation s. Together these are called the **summary statistics** of the set. Here are their formulas.

$$\overline{X} = \frac{1}{n}\sum_{i=1}^{n} x_i \qquad s = \left(\frac{\sum_{i=1}^{n}(x_i - x)^2}{n - 1}\right)^{1/2}$$

Two sets of data can differ in their average values, or they can differ in their standard deviations. The scientists might judge either of these differences to be noteworthy.

Statisticians call this **significance testing,** and they have developed tests to judge the significance of a difference between summary statistics. What do they mean by a **"significant difference"**? A difference that is not the result of random error. Unfortunately it is not possible to determine **with certainty** whether a particular difference is caused by random error; however, statisticians have been able to develop a procedure that can **compute this probability** (based on certain assumptions). If we know the probability that the difference could be caused by random error, then we also know the probability that it is not caused by random error, so we will know the probability of a "significance difference." That is the best that the statistical test can do.

At this point, experimental scientists have to make a decision: how much leftover "random" probability are they willing to live with? With a brand-new result, it is unlikely that their initial measurements are decisive; the scientists will be betting that their limited set of data has really uncovered something "significant." Their professional reputations—perhaps even their jobs—could be at stake, so they will want to know the probability associated with their bet! In chemistry and physics, an experimental difference is usually labeled "significant" when the probability that it is not due to random error is **greater than 95%.** In other fields the threshold probability may be higher or lower than 95%. The threshold is never determined by statisticians; it is determined collectively by the experimentalists who work in a particular field.

There are two tests designed to test differences among summary statistics: the **t test** tests the difference of two averages, and the **F test** tests the difference of two standard deviations. This activity will examine the *t* test. Because there are three different variations of the *t* test, this activity is rather long, and so it has been divided into two parts. *Part 1* develops an understanding of how the *t* test does its job, and *Part 2* examines the three different types of *t* tests and how they can be performed in Excel. Another activity covers the *F* test.

Consider this...

You put 19 white balls and 1 red ball, all the same size, into a large, black bag. You can't see the balls in the bag, but you can reach into the bag and hold them.

Key Questions

1. You reach into the bag, grab one ball, and remove it. What is the probability that you will withdraw a white ball?

2. If you could find someone who would take the bet, would you bet $50 that you could withdraw a white ball from the bag? (You can assume that you have $50.) Compare your answer with those of your group.

3. For group discussion: Do you think that 95% probability is a reasonable threshold for accepting a new claim in chemistry or physics?

4. What areas of experimental science or engineering should have a threshold probability much higher than 95%? Record your group's best answer here.

Consider this...

April and Betty each measure the absorbance of the same colored solution three times; each student has her own absorption spectrometer. (Absorbance is unitless.)

Set 1	
April	**Betty**
0.121	0.123
0.122	0.124
0.123	0.125
$\bar{X}$ 0.122	0.124
s 0.001	0.001

Key Questions

5. Compute the summary statistics for April and Betty's data and enter them in the table.

> The symbol P_{signif} represents the probability that the difference between April's data and Betty's data is significant, that is, that the difference is the result of some real difference in the sample or the measuring routine. $P_{insignif}$ will represent the probability that the difference is the result of random error.
>
> (Note: Statisticians often use the symbol α for $P_{insignif}$; $\alpha=0.05$ means $P_{signif}=0.95$)

6. What is the sum of P_{signif} and $P_{insignif}$? Why?

T-test is $P_{insignif}$ = how probable is it that the 2 tests diffes due to error

Consider possible claims regarding April and Betty's data:

a. Betty's absorption data are higher than April's; therefore Betty's spectrometer must somehow be different from April's, making it give higher measurements than April's.

b. April's absorption data are lower than Betty's; therefore April must have somehow diluted her sample slightly before she made her measurements.

c. Betty and April's absorption data are a little different, but not by much, so maybe the difference is entirely random.

d. There's no real difference between April's and Betty's data; the difference is within the limit of performance of the two spectrometers.

---

7. Which claim(s) say that the difference in April and Betty's data is **insignificant**?

    Which claim(s) says that the difference in their data is **significant** and "real?"

8. If two people claim that a difference is real, do they have to agree on the cause of the difference? Explain.

9. Assume that P_{signif} = 0.88 for April and Betty's data. Would we judge the difference to be significance or insignificant? Which claims would be ruled out?

10. Assume that P_{signif} = 0.97. Would we judge the difference to be significant or insignificant? Which claims would be ruled out?

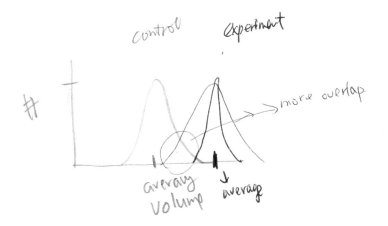

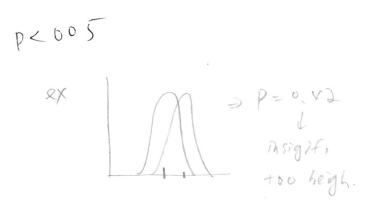

Consider the following possible sets of data that April and Betty might obtain. Set 1 is listed again for easy comparison. Determine the means and standard deviations by inspection. Divide the work up among your group.

Set 1	
April	**Betty**
0.121	0.123
0.122	0.124
0.123	0.125
$\overline{X}$	
S	

Set 2	
April	**Betty**
0.120	0.124
0.121	0.125
0.122	0.126
$\overline{X}$	
S	

Set 3	
April	**Betty**
0.121	0.121
0.122	0.122
0.123	0.123
$\overline{X}$	
S	

Set 4	
April	**Betty**
0.120	0.122
0.122	0.124
0.124	0.126
$\overline{X}$	
S	

Set 5	
April	**Betty**
0.122	0.124
0.122	0.124
0.122	0.124
$\overline{X}$	
S	

Set 6	
April	**Betty**
0.122	0.124
$\overline{X}$	
S	

Set 7	
April	**Betty**
0.121	0.123
0.121	0.123
0.122	0.124
0.122	0.124
0.123	0.125
0.123	0.125
$\overline{X}$	
S	

Changing the difference in the means

11. a. Consider Set 2. **Compared to Set 1**, which summary statistics—the average, the standard deviation, or both—have changed?

Compared to Set 1, what has happened to the value of P_{signif}? Has it increased or decreased? Explain your conclusion.

b. Consider Set 3. **Compared to Set 1**, which summary statistics have changed?

Compared to Set 1, what has happened to the value of P_{signif}? Explain your conclusion.

c. Based on your answers to the previous two questions, what should happen to P_{signif} **as the difference between the average values increases**? Your answer should be a complete, grammatically correct sentence.

Changing the standard deviations

12. a. Consider Set 4. **Compared to Set 1**, which summary statistics have changed?

Compared to Set 1, what has happened to the value of P_{signif}? Explain your conclusion.

b. Consider Set 5. **Compared to Set 1**, which summary statistics have changed?

Compared to Set 1, what has happened to the value of P_{signif}? Explain your conclusion.

c. Based on your answers to the previous two questions, what should happen to P_{signif} **as the standard deviation of the measurements increases**? Your answer should be a complete, grammatically correct sentence.

Changing the number of measurements

13. a. Suppose that April and Betty had made just one measurement on the sample, as in Set 6. **Compared to Set 1**, what has happened to the value of P_{signif}? Explain your conclusion.

b. Suppose that April and Betty each made six measurements, as in set 7. **Compared to Set 1**, what has happened to the value of P_{signif}? Explain your conclusion.

c. Based on your answers to the previous two questions, what should happen to P_{signif} **as the number of measurements increases**? Your answer should be a complete, grammatically correct sentence.

14. Develop a group summary of the effect that the standard deviations, the number of data, and the difference in the averages have on P_{signif}.

15. Discuss as a group: If you had to bet to $50, which set of data, 1–7, would be the best bet to have a significant difference in the average values? Explain your choice.

Consider this...

The difference between average values of two sets of measurements is a quantity derived from the data, and as such it has an error, its limit of **precision**. The error can be computed using the rules for the propagation of error. Each step in the calculation requires a corresponding step of error propagation.

Step 1: *Computing the averages and their errors.*

The standard deviation of an average value is $s/\sqrt{n}$; where s is the standard deviation of the measurements and n is the number of measurements in the set. ($s/\sqrt{n}$ is sometimes called the **standard error** or the **standard deviation in the mean**. Throughout the remainder of this activity, the former name will be used.) If we have two data sets, then we get the average $\bar{x}_1$ and its error $S_{\bar{x}_1}$, and x_2 and its error $S_{\bar{x}_2}$, where

Standard error: $S_{x_i} = \dfrac{S_i}{\sqrt{n_i}}$

Step 2: *Computing the difference and its error:*

The error in the difference of two measurements is $\sqrt{S_1^2 + S_2^2}$, the square root of the **sum** of the squared errors. When we take the difference of two average values, $\Delta\bar{x} = \bar{x}_1 - \bar{x}_2$,

and the **Standard error of** $\Delta\bar{x}$ is $S_{\bar{x}_1 - \bar{x}_2} \equiv S_{\Delta\bar{x}} = \sqrt{S_{\bar{x}}^2 + S_{\bar{x}}^2}$.

The error in the difference of the averages is the square root of the sum of the squares of the **standard errors.** Using the original standard deviations of the measurements, the equation for the standard error in the difference is

$$S_{\Delta\bar{x}} = \sqrt{\dfrac{S_1^2}{n_1} + \dfrac{S_2^2}{n_2}}$$

Relative error is the error in a quantity divided by the quantity itself. We're interested in the relative error of the differences in the averages—well, actually we're interested in its reciprocal. $\Delta\bar{x}$ is the difference in the means, and $S_{\Delta\bar{x}}$ is its error. We can put these two pieces of information together in one formula.

Reciprocal of relative error: $\dfrac{\Delta\bar{X}}{S_{\Delta\bar{x}}} = \dfrac{\Delta\bar{X}}{\sqrt{\dfrac{S_1^2}{n_1} + \dfrac{S_2^2}{n_2}}}$

---

16. Develop a group summary: Does the mathematical expression for the reciprocal of the relative error in the difference **change as you expect P_{signif} to change** for different sets of data? Your summary should explain how the quantity will change as the standard deviations of the measurements increase, how it will change as the number of measurements increases, and how will it change as the difference of the average values increases.

Consider this...

The reciprocal of relative error is what analytical chemists generally call the *t* **value**; *t* is defined as the absolute value of the difference of the means divided by the error in the difference. **A *t* value is simply the unsigned difference divided by its error.** A *t* value is also the reciprocal of relative error.

Equations for *t*

$$t = \frac{|\Delta \overline{X}|}{S_{\Delta \overline{x}}} = \frac{|\overline{X}_1 - \overline{X}_2|}{\sqrt{S_{\overline{x}_1}^2 + S_{\overline{x}_2}^2}} = \frac{|\overline{X}_1 - \overline{X}_2|}{\sqrt{\dfrac{S_1^2}{n_1} + \dfrac{S_2^2}{n_2}}}$$

Key Questions

17. If a particular difference $\Delta \overline{X}$ has **10% relative error,** what is the value of *t*?

18. Why do we use the absolute value? Why can we disregard the sign of the difference?

19. As a group, verify that for every set of April and Betty's data, $n_1 = n_2$ and $s_1 = s_2$. Under these circumstances, the formula for t can be simplified. Letting $s_1 = s_2 = s$ and $n_1 = n_2 = n$, show that the formula becomes

$$t = \frac{|\bar{X}_1 - \bar{X}_2|}{\sqrt{\dfrac{s_1^2}{n_1} + \dfrac{s_2^2}{n_2}}} = \frac{|\bar{X}_1 - \bar{X}_2|}{\sqrt{\dfrac{2s^2}{n}}} = \frac{|\bar{X}_1 - \bar{X}_2|}{s\sqrt{\dfrac{2}{n}}}$$

20. Using the simplified formula in Q19, fill in the following table for each set of data. Divide up the work among the members of your group.

Set	$|\bar{X}_1 - \bar{X}_2|$	s	n	t
1	0.002	0.001	3	2.449
2				
3				
4		0.002		
5				
6	0.002	Unknown	1	NA
7		0.0009	6	3.873

21. There are two cases where it is not possible to compute t. What problems prevent you from computing t?

22. Discuss as a group: Does t change as you expected P_{signif} to change in these data sets? Do t and P_{signif} have the same trend as the data change? Give two examples supporting your answer and explain your reasoning.

23. Discuss as a group: What is the range of possible numeric values for t? (That is, what values could it take on?)

What is the range of possible numeric values for P_{signif}?

24. Support or refute the following statement: t is the same numerical quantity as P_{signif}.

Consider this...

It is possible to use the t value to determine P_{signif}—but right now all we need to do is find out if $P_{signif} \geq 0.95$. There is a table of t values that tells us when t is big enough to know that $P_{signif} \geq 0.95$. The table is given below. It is a portion of the **Student's t table for two-tailed distributions.** (Excel's TINV function will provide these values.)

Degrees of freedom	$t_{critical}$ for $P_{sig} \geq 0.95$
1	12.706
2	4.303
3	3.182
4	2.776
5	2.571
6	2.447
7	2.365
8	2.306
9	2.262
10	2.228
15	2.131
20	2.086
25	2.060
30	2.042
40	2.021
50	2.009
60	2.000
120	1.980
∞	1.960

When you compare a t value for experimental data against a $t_{critical}$ value, you are doing a **t test**. In different textbooks the t value computed from the data may be called $t_{calculated}$ or t_{calc} or t_{data} or t_{stat} or simply t; the t value from the Student's t table is usually called $t_{critcal}$ or t_{table}. There are several types of t tests, but in all cases when $t_{calculated} > t_{critical}$, the difference being tested is considered significant.

For the rest of *Part 1* of this activity, and also in *Part 2*, we will use the symbols **t_{calc}** and **t_{crit}**. t_{calc} is computed from the data; t_{crit} comes from the table. When the $t_{calc} > t_{crit}$, we know that $P_{signif} > 0.95$, and **we can conclude that the difference in the means is real and significant.**

Key Questions

25. What would we conclude about P_{signif} if t_{calc} equaled t_{crit}?

26. If a particular difference $\Delta \overline{X}$ has **10% relative error,** is the difference probably significant? Explain.

27. To find a particular t_{crit} value in the Student's t table, what quantity do you need?

The number of **degrees of freedom** is the total number of data minus the number of averages that were computed from the data.

Degrees of freedom: $n_{df} = n - p$

where p is the number of averages that were computed.

---

28. If April and Betty have each collected 3 data points, and they each compute 1 average value from their data, they each have 2 degrees of freedom.
What is the total degrees of freedom from both sets of data?

How many degrees of freedom are there when April and Betty each collect 6 data points?

29. Complete the following table for each set of data. Enter t from your previous table as t_{calc}. Divide the work among the members of your group.

Set	t_{calc}	n_{df}	t_{crit}	Significant?
1	2.449			no
2		4		
3				
4				
5				
6	NA	0		
7	3.873		2.228	YES!

30. Discuss in your group: Among the five cases where it was possible to compare t_{calc} and t_{crit}, which ones showed a significant difference? Are there any cases where you and your group are surprised that the P_{signif} is greater than 0.95?

31. Review your answer to Q15. Which set appears to have the **most** significant difference? What makes you think that it is most significant?

What is it about this data set that leads to this result?

Would anyone in your group bet $50 that this difference is significant?

32. What question about data sets does the t test answer?

33. For group discussion: On a scale from 1 to 10, how confident are you that you understand how the t test is used?

34. On a scale from 1 to 10, how confident are you that you can find a required value in the Student's t table?

Applications

35. Consider the following statement: "Extraordinary claims require extraordinary evidence." How can the t test be used to assess extraordinary claims?

36. Consider the following statement: "Significant differences deserve an explanation; differences that are not significant do not!" What role does the t test play in this statement?

37. Consider the two cases where a t test could not be completed. How are the two cases different? Based on the behavior of t_{crit} in the Student's t table, what conclusions can you draw about the significance of the differences in these two cases?

38. Consider Set 1. Generate a set of three data that is significantly different from April's data but not significantly different from Betty's data.

39. Suppose you had two data sets each containing four data, and the two sets have equal standard errors. How far apart would the two means have to be, *in terms of the standard error,* for the t test to yield a significant difference?

40. Blood is drawn from a patient suffering from hypoparathyroidism to determine her serum calcium concentration. Five replicate analyses yield a mean of 8.11 mg/dL with a standard error of 0.16 mg/dL. Two weeks later after drug therapy her serum calcium is measured again; now five measurements yield 8.40 mg/dL with a standard error of 0.19 mg/dL. Has the patient's serum calcium changed?

41. Sally measures the amount of CO_2 in a pint soda bottle by weight loss; she shakes the bottle vigorously and then *slowly* opens the cap, letting the gas escape. Three measurements on Brand X indicate a loss of 2.2 ± 0.2 g of gas, while three measurements on Brand Y show a loss of 2.4 ± 0.3 g. (The ± indicates standard error.) Do Brands X and Y have different amount of CO2?

42. Suppose that Bob and Bill make repeated measurements of the amount of iron in two iron ore samples. Bob gets 20.14%, 20.14%, 20.20%, and 20.26% Fe for his ore sample, and Bill gets 20.02%, 20.08%, 20.11%, and 20.18% Fe for his sample. Do the two samples have different iron percentages?

Part 2: Excel Implementation and Variations

The t test that we have been considering is called a **Comparison of Two Means,** or a **Comparison of Replicate Measurements.** This test tells us whether $\Delta\bar{x}$, the difference in the average value of two sets of measurements, is statistically significant.

Review: In *Part 1* we learned that the value t_{calc} is the ratio of the difference of the means to the standard error in that difference.

Comparison of Two Means

$$t_{calc} = \frac{|\Delta\bar{X}|}{S_{\Delta\bar{x}}} = \frac{|\bar{X}_1 - \bar{X}_2|}{\sqrt{\dfrac{S_1^2}{n_1} + \dfrac{S_2^2}{n_2}}}$$

As t_{calc} increases, the probability of a significant difference increases.

We saw that t_{calc} increases with increasing difference in the means, increasing numbers of data, and decreasing standard deviations. To do a t test, t_{calc} is compared with t_{crit}, a value that is selected from the Student's t table based on the number of degrees of freedom n_{df}. If $t_{calc} > t_{crit}$, then the probability of a significant difference, P_{signif}, is greater than 0.95 and the difference is considered significant.

TTtest (array 1, array 2, tails, paired/unpaired)
         (data 1)   (data 2)

Consider this…

paired equal variance
1.

unpaired equal variance
2 or 3

In the examples we examined in Part 1, April and Betty always had the same number of data, and the standard deviations of their measurements were always equal. Clearly this will not always be the case. In set 2 below April and Betty have different numbers of data and different standard deviations. The mean and variance are listed for each group of data. (The variance v is the square of the standard deviation.)

→ need F-test

⟶ F-test
unpaired

	Set 1		Set 2		Set 3	
	April	Betty	April	Betty	April	Betty
	0.121	0.123	0.121	0.123	0.121	0.123
	0.122	0.124	0.122	0.123	0.121	0.123
	0.123	0.125	0.123	0.124	0.122	0.124
			difference 0.124	0.124	0.122	0.124
				0.125	0.123	0.125
				0.125	0.123	0.125
$\bar{X}$	0.122	0124	0.122	0.124	0.122	0.124
v	10×10^{-07}	10×10^{-07}	10×10^{-07}	8×10^{-07}	8×10^{-07}	8×10^{-07}

variance

variance

TT est   0.070484   0.01845   0.00309

① (data 1, data 2, 2, 2)

② (data 1, data 2, 1, 2) 0.03552 (# less)

Key Questions

1. Consider t_{calc} for each of these data sets. Rank the sets in the order that you expect t_{calc} to fall, from lowest to highest, and explain your prediction.

2. Test your predicted order. For each set of data, compute t_{calc}.
   Divide the work among your group. Was your predicted order correct?

3. Determine t_{crit} for set 2. Is the difference between April's and Betty's means significant in this case?

Here are the results of a t test performed on Set 2 in **Excel** using the *t-test: Two-Sample Assuming Equal Variances* tool in the Data Analysis Toolpak. The default values of the test were used.

	April	Betty
Mean	0.122	0.124
Variance	0.000001	8E-07
Observations	3	6
Pooled Variance	8.57E-07	
Hypothesized Mean Difference	0	
Df	7	
t Stat	-3.05505	
P(T<=t) one-tail	0.009226	
t Critical one-tail	1.894579	
P(T<=t) two-tail	0.018452	
t Critical two-tail	2.364624	

*This Excel test reports a negative value of t. In the Excel input window, if the columns containing April and Betty's data are swapped, the t test tool will report a t Stat equal to +3.055. For this type of t test, you can ignore a negative sign in a t value from Excel.

4. As a group, examine the t test results from Excel. Based solely on the Excel results, is the difference between April's and Betty's means significant? Which values in the output did you use to make your decision?

Here are the essential test results from Excel for all three sets. (Negative signs removed)

	Set 1	Set 2	Set 3
Pooled Variance	0.000001	8.57E-07	8E-07
df	4	7	10
t Stat	2.44949	3.05505	3.87298
P(T<=t) two-tail	0.070484	0.018452	0.003094
t Critical two-tail	2.776445	2.364624	2.228139

5. Compare the results from Excel with your results in Questions 2 and 3. Check t_{calc}, n_{df}, and t_{crit}. Can you assign these values to the Excel output? Are there any values that disagree with your results?

You should find that your value of t_{calc} for Set 2 is slightly less than the "t Stat" value from Excel (though it doesn't affect the decision that the means are significantly different.) However, your t_{calc} for Sets 1 and 3 agree with Excel.

6. Compare the quantities labeled Pooled Variance with the variances listed for each group of data. What conclusion can you draw about the value of the pooled variance when the individual variances are equal?

The **pooled standard deviation** is a way to combine standard deviations for two sets of measurements if the variances *are essentially equal*. It simplifies the formula for the standard error of the difference.

$$S_{pooled} = \sqrt{\frac{(n_1 - 1)s_1^2 + (n_2 - 1)s_2^2}{n_1 + n_2 - 2}}$$

The equation for the pooled variance is $v_{pooled} = S^2_{pooled}$

Using the pooled standard deviation, we can express $S_{\Delta \bar{x}}$ more simply:

$$S_{\Delta \bar{x}} = S_{pooled} \sqrt{\frac{1}{n_1} + \frac{1}{n_2}}$$

7. Show that, using the pooled standard deviation, the formula for t_{calc} becomes

$$t_{calc} = \frac{|\bar{X}_1 - \bar{X}_2|}{S_{pooled}} \sqrt{\frac{n_1 n_2}{n_1 + n_2}}$$

8. The t test used in Excel made the assumption that the variances of the two sets of data were equal. Are the two variances in Set 2 equal?

In another activity, we will introduce a test called the F test that will determine if the variances are **significantly** different. There are actually two forms of the t test, and the choice of t test is based on the result of the F test. If the variances are not significantly different, then t is computed using the formula in Question 8 using s_{pooled}. If the variances are significantly different, then s_{pooled} is not used, and t_{calc} is computed using the boxed formula at the beginning of Part 2. But there are other differences as well; these will be covered in the F test activity.

In theory, we should always do an F test before we do the t test—but in practice the F test is often skipped. In routine chemical analysis on small data sets that come from the same measuring device, it is reasonable to assume that the variances are not significantly different, even if they are not exactly equal. But it is always safer to test that assumption than to proceed in error.

9. In the Excel output, there is a quantity labeled P(T<=t) two-tail. This quantity is a probability. Is it P_{signif} or $P_{insignif}$? Explain your choice.

> * One of the default values for the t test is "alpha=0.05". Statisticians use α to represent $P_{insignif}$; alpha = 0.05 makes the t test use a t_{crit} consistent with a 95% probability of a real difference. Statisticians usually base their tables on $P_{insignif}$, so be careful when you are looking at P values—If you see an α, assume the α probabilities are $P_{insignif}$.

10. For group discussion: If you *had to bet* $50 on something, what would be the better bet: that the difference in Set 2 is real, or that you will draw one card from a standard pack of 52 cards and **not** draw a deuce? Explain.

11. As a group, write a set of instructions that would let a lab technician read the output of an Excel t test and come to a conclusion about significance.

Consider this...

The most common type of t test is the Comparison of Means, but there are two other important applications of the t test. Both can be considered special cases of the Comparison of Means.

The first is the **Comparison against an Accepted Value.** Often we want to compare one set of measurements against a value that is "known" or accepted as "true." The true value is widely regarded as "above reproach" and can be assumed to have no error.

12. For group discussion: If the accepted value has no error, what is its "standard deviation?"

13. Assuming that $\bar{X}_2$ = True and $S_{True} = 0$, show that the formula for t_{calc} for the Comparison Against an Accepted Value becomes

$$t_{calc} = \frac{|\bar{X}_1 - True|}{S_{\bar{x}_1}} = \frac{|\bar{X}_1 - True|}{S_1}\sqrt{n}$$

Comparison against an Accepted Value	$t_{calc} = \dfrac{|\bar{X}_1 - True|}{S}\sqrt{n}$

14. April and Betty are told that the accepted value for the absorption of their solution is 0.122. Using the data in Set 2, determine whether April and Betty's average values are significantly different from 0.122.

Consider this...

Another type of t test is the **Comparison of Paired Values** or **Comparison of Individual Differences**. It is used to test the significance of a series of *differences* that result from paired measurements on the same sample.

Suppose that April and Betty get six **different** colored **solutions** and measure the absorptions of the solutions using their instruments. They now have six **pairs of measurements**, and they can compute six differences, one difference for each pair: $d_1 = A_{Betty1} - A_{April1}$, $d_2 = A_{B2} - A_{A2}$, d_3, d_4, d_5, and d_6. From these values of d they compute the average difference $\bar{d}$, and s_d, the standard deviation of the differences.

15. For group discussion: April and Betty hope that their measurements agree. What values would they like to get for $\bar{d}$ and s_d?

16. Realistically, since there is always some random error present in any measurement, April and Betty will not get $s_d = 0$. If their measurements mostly agree, should $\bar{d} > s_d$ or should $\bar{d} < s_d$? Explain your group's reasoning.

17. April and Betty need to test $\bar{d}$ to see if it is significantly different from 0. (Even though they're hoping that it won't be!) Show that if $\bar{X} = \bar{d}$ and *True*=0, the formula for t_{calc} for the Comparison of Paired Values becomes

$$t_{calc} = \frac{|\bar{d}|}{S_{\bar{d}}} \quad \text{or simply}$$

Comparison of Paired Values $\quad t_{calc} = \dfrac{|\bar{d}|}{S_d} \sqrt{n}$

18. Since April and Betty are testing an average based on six values (even though there were originally 12 measurements), for their paired t test, $n=6$. So what is n_{df}?

19. a. Assume that April and Betty get identical measurements for three solutions and get a difference of +0.001 for the other three. (Betty's values are always higher.)
Generate a set of data consistent with the assumption.

b. Obviously $\bar{d} = +0.0005$. Find $S_{\bar{d}}$. Is it > or < $\bar{d}$?

c. Compute t_{calc}.

d. Find t_{crit}. Are April and Betty getting significantly different measurements? Explain your conclusion in a complete sentence.

20. Most Student's t tables have columns of t_{crit} values for other values of P_{signif}. Usually these probabilities are called "confidence levels" and are given in percentages. For example, for 5 degrees of freedom, you would find the following values:

n_{df}	50%	90%	95%	98%	99%	99.5%	99.9%
5	0.727	2.015	2.571	3.365	4.032	4.773	6.869

Based on the t_{crit} values for $n_{df}=5$ and the t_{calc} you computed in Question 19, bracket the value of P_{signif}:

_____ < P_{signif} < _____

21. Here is the output from Excel's Data Analysis ToolPak for a *t test: Paired Two Sample for Means*. The input data was consistent with Q19.

	April	Betty
Mean	0.158	0.1585
Variance	0.017432	0.01745
Observations	6	6
Pearson Correlation	0.999992	
Hypothesized Mean Difference	0	
df	5	
t Stat	-2.23607	
P(T<=t) one-tail	0.037793	
t Critical one-tail	2.015049	
P(T<=t) two-tail	0.075587	
t Critical two-tail	2.570578	

Since April's data are always less than or equal to Betty's and April's data were placed in the first column, the value of t Stat is negative. Otherwise, is it consistent with your value of t_{calc}? Explain.

What other values do you recognize in the Excel output? Do they match your values?

22. Excel reports that P(T<=t) two-tail is 0.076. What does this indicate about P_{signif} and $P_{insignif}$? Is it consistent with your conclusion about P_{signif} in Q20?

23. As a group, review your set of instructions for a lab technician in Q11. Would they work for the Paired Two-Sample *t* test? If not, modify them accordingly and write them below.

24. What are the three types of *t* tests, and what questions do they answer?

25. For group discussion: How confident are you that you will know which *t* test to use in a given situation?

26. On a scale from 0 to 10, rate your ability to read Excel output.

Applications

27. Insert the data from Set 2 into Excel and verify that you get the Excel output shown in Q3.

28. Suppose you had a set of four data points with a certain mean and standard error. Suppose there is an accepted value for these data. *In terms of the standard error,* how far away from the accepted value would the mean have to be for the t test to determine a significant difference?

29. Suppose that Bob and Bill make repeated measurements of the amount of iron in two iron ore samples. Bob gets 20.14%, 20.14%, 20.20%, and 20.26% Fe for his ore sample, and Bill gets 20.02%, 20.08%, 20.11%, and 20.18% Fe for his sample. Bob and Bill learn that their samples are from the same source, and that the accepted value is 20.09% Fe. Are Bob's and Bill's means significantly different from the accepted value?

30. Suppose you have a set of seven paired samples with an average difference of $\bar{d}$ and a standard error in the difference of $S_{\bar{d}}$. *In terms of the standard error,* how big must $\bar{d}$ be for the paired t test to determine a significant difference?

31. Insert the data you generated in Q19 into Excel, perform the Paired Two-Sample t test, and show that you get the results displayed in Q21.

32. Eight patients are treated for hypoparathyroidism using a drug therapy. Before the treatment, the eight patients had serum calcium levels of 8.12, 7.98, 8.06, 7.65, 8.20, 8.08, 8.15, and 8.22 mg/dL. After treatment, their serum levels were 8.22, 8.02, 8.08, 7.96, 8.31, 8.28, 8.20, and 8.33 mg/dL. Was there a significant increase in serum calcium after the patients were given the treatment? Which t test should be used to answer this question?

Statistical Tests of Data: The *F* Test

Learning Objectives

Students should be able to:

Content

- Explain how the *F* value changes with respect to the standard deviations of the data sets.

- Explain how the *F* test is used to make decisions regarding significance, and be able to use Excel's *F* test output to make such decisions.

- Explain how the results of the *F* test will affect the *t* test.

Process

- Interpret tables of data. (Information processing)

- Draw conclusions about data sets. (Critical thinking)

- Express concepts in grammatically correct sentences. (Communication)

Prior knowledge

- Familiarity with common statistics terms such as mean, range, standard deviation, and variance.

- The *t* test activity.

Further Reading

- Harris, D.C. 2007. *Quantitative Chemical Analysis,* 7th Edition, pp. 59-65. New York: WH Freeman.

- Hecht, H.G. 1990. *Mathematics in Chemistry: An Introduction to Modern Methods,* pp. 210-255. Englewood Cliffs, NJ: Prentice Hall.

- NIST website. http://www.itl.nist.gov/div898/handbook/prc/section3/prc31.htm (accessed July 2008).

Author

Carl Salter

PGLANA004012a POGIL

Consider this...

Scientists often compare sets of similar measurements. For example, an analytical chemist might compare two different methods for measuring the amount of carbon dioxide in air. Or a food chemist might measure dissolved carbon dioxide in two different brands of soda. Even when the comparison isn't explicit, analysts are often implicitly comparing their measurements against reference samples or accepted standards. Whatever the case, experimental scientists often must decide if an observed **difference** is important and worthy of communication. Usually they will focus on the two most important statistics about a data set: its average $\bar{X}$ (or mean) and its standard deviation s. Together these are called the **summary statistics** of the set. Two sets of data can differ in their average values, or they can differ in their standard deviations—the scientists might judge either of these differences to be noteworthy.

As a young scientist, you probably realize why it is important to compare average values, but you may not understand why anyone would need to compare standard deviations. Let's extend an example mentioned above: Suppose that an analytical chemist found that two brands of soda had the same average dissolved CO_2 concentration, but that the standard deviations were fairly different. Let's say that Brand X has a standard deviation that is two times that of Brand Y. On average the same amount of gas is getting into both brands of soda, but something is increasing the range of values in Brand X. The bigger standard deviation could mean that Brand X's process for dissolving gas in its soda is not as carefully controlled as Brand Y's. But before we can draw this conclusion and launch an expensive investigation, we have to know whether a factor of two difference in standard deviation is **significant;** that is, probably not the result of random error in the measurements.

Unfortunately it is not possible to determine **with certainty** whether a particular difference is caused by random error; however, statisticians have been able to develop a procedure that can **compute this probability** (based on certain assumptions). If we know the probability that the difference could be caused by random error, then we also know the probability that it is not caused by random error, so we will know the probability of a "significance difference." That is the best that the statistical test can do.

In chemistry and physics an experimental difference is usually labeled "significant" when the probability that it is not due to random error is **greater than 95%.** (In other fields the threshold probability may be higher or lower than 95%.) So if there is at least a 95% chance that the difference in standard deviations between Brands X and Y is not due to random error, we would conclude that the difference is real and needs to be explained.

This activity examines the **F test,** which tests the difference of two standard deviations for significance. *Part 1* of this activity develops an understanding of how the F test does its job. *Part 2* examines how the F test affects the t test, which tests average values for significant differences. The results of an F test for standard deviations will dictate how a subsequent t test on the same data must be performed. Both parts of this activity require some previous experience with the t test.

Part 1: Fundamentals of the *F* test

Consider this...

Comparing average values involves a quantity t, and comparing t_{calc} for two data sets with a $t_{critical}$ value is called a **t test**. In the same way, comparing standard deviations, or variances (the square of the standard deviation) involves a quantity called F, and comparing F_{calc} for two data sets with an $F_{critical}$ value is called an **F test** (or Fisher test). It is easiest to define F_{calc} using variances. (Variance is given the symbol ν, the Greek letter nu.)

Consider the following data, their summary statistics and variances:

	April	**Betty**	**Charly**	**David**
	0.121	0.123	12.1	12.3
	0.122	0.123	12.2	12.3
	0.123	0.124	12.3	12.4
		0.124		12.4
		0.125		12.5
		0.125		12.5
$\bar{x}$	0.122	0.124	12.2	12.4
s	1.0×10^{-3}	0.89×10^{-3}	0.1	0.089
ν	10×10^{-7}	8×10^{-7}	0.01	0.008

ftest

Key Questions

1. The t_{calc} value is a unitless ratio. For a Comparison of Means t test, Charly and David will get the same value for t_{calc} that April and Betty get. Why?

2. Examine the variances of the data. There is some probability that April and Betty's data have significantly different variances. There is some probability that Charly and David's data have significantly different variances. Compare those two probabilities; are they the same or different? Explain your conclusion in a grammatically correct sentence.

F_{calc} is also a unitless ratio; it is simply the ratio of the variances for the two sets of data.

$$F_{calc} = \frac{v_1}{v_2}$$

3. What is the value of F_{calc} for April and Betty's data?

4. Compare the F_{calc} value for April and Betty's data with that for Charly and David's data. Are they the same or different? Is this consistent with your answer in Q2?

5. Complete the following equation:

$$F_{calc} = \frac{v_1}{v_2} = \frac{s_1^2}{s_2^2} = \left(\frac{\quad}{\quad} \right)^2$$

6. Based on your answer to Q5, write a grammatically correct sentence that describes F_{calc} in terms of the standard deviations.

7. What value of F_{calc} would indicate that the variances are <u>not different?</u>

Consider this...

Examine the following portion of the F_{crit} table for a 95% probability of a significant difference.

| | $F_{critical}$ Table for 95% Significance | | | | | |
| | n_{df} numerator | | | | | |
n_{df} denominator	1	2	3	4	5	6
1	161.45	199.50	215.71	224.58	230.16	233.99
2	18.51	19.00	19.16	19.25	19.30	19.33
3	10.13	9.55	9.28	9.12	9.01	8.94
4	7.71	6.94	6.59	6.39	6.26	6.16
5	6.61	5.79	5.41	5.19	5.05	4.95
6	5.99	5.14	4.76	4.53	4.39	4.28

Key Questions

8. What are the row and column indexes in the F table? How is n_{df} related to the number of data in a given set?

9. Based on the values in the F table, which variance must go in the numerator when you compute F_{crit}?

By convention, the larger variance goes in the numerator, so F values are always greater than one, just as you see in the table above.

10. For group discussion: What inequality relationship involving F_{calc} and F_{crit} indicates that the difference in variances is significant? Explain your reasoning.

11. Suppose you have two sets of data each containing five values. Based on the F table, what ratio of standard deviations would represent a significant difference? Compare your results with the rest of the group.

12. What is n_{df} for April's data? For Betty's data? For a 95% probability of a significant difference in the variances, what is F_{crit} for their data?

13. Are April and Betty's variances significantly different? Explain your conclusion.

Consider this...

Here are the results of an *F* test performed in Excel using the *F-Test Two-Sample for Variances* tool in the Data Analysis Toolpak. The default values of the test were used.

	April	Betty
Mean	0.122	0.124
Variance	0.000001	8E-07
Observations	3	6
df	2	5
F	1.25	
P(F<=f) one-tail	0.362887	
F Critical one-tail	5.786135	

Key Questions

14. Examine the Excel output. Circle the values you recognize. Do those values agree with your previous results?

15. As a group, discuss the significance of the probability called P(F<=f) one-tail in the Excel output. (Don't worry about the meaning of "one-tail" or "F<=f"; focus on the numerical value.) What does the value tell you about the probability of a significant difference in the variances?

16. As a group, write a set of instructions that would let a lab technician read the output of an Excel *F* test and come to a decision on significance.

17. Create a new data set for April that will have a significantly different variance from Betty's data, according to the *F* test. Do this without changing the average value of April's data, and without changing the number of data.

Part 2: How the F test result affects the t test

Consider this...

	Evan	Fred
	12.0	12.0
	12.2	12.1
	12.3	12.2
	12.6	12.2
	12.7	12.2
	12.8	12.3
		12.4
$\bar{x}$	12.43333	12.2
v	0.098667	0.016667

Ftest (data1, data2) = 0.00513,

Ttest (P-value) unpaired = 0.009835
equal

Key Questions
Ttest (P-value) paired(1) = 0.00~0.1365
un equal variance

18. Perform an *F* test on Evan and Fred's data. Are their variances significantly different? Explain your reasoning.

{ paired
unpaired equal variance
unpaired unequal variance.

Examine the following Excel output based on Fred and Evan's data:

t Test Assuming Unequal Variances	Evan	Fred
Mean	12.43333	12.2
Variance	0.098667	0.016667
Observations	6	7
Hypothesized Mean Difference	0	
df	6	
t Stat	1.70061	
P(T<=t) one-tail	0.069961	
t Critical one-tail	1.943181	
P(T<=t) two-tail	0.139922	
t Critical two-tail	2.446914	

t Test Assuming Equal Variances	Evan	Fred
Mean	12.43333	12.2
Variance	0.098667	0.016667
Observations	6	7
Pooled Variance	0.053939	
Hypothesized Mean Difference	0	
df	11	
t Stat	1.805829	
P(T<=t) one-tail	0.049177	
t Critical one-tail	1.795884	
P(T<=t) two-tail	0.098354	
t Critical two-tail	2.200986	

19. Based on the results of the F test, which t test should be used?

20. What conclusion would you draw about the difference of Evan's and Fred's means based on the *Unequal Variances* output? Based on the *Equal Variances* output? (Reminder: to compare means using Excel output, use t Stat and t Critical two-tail.)

21. Consult a t table. Are the t_{crit} values consistent with the number of degrees of freedom in each table?

22. Discuss as a group: What is the most surprising difference between the two tables, and why?

When **the variances are not equal,** the degrees of freedom is computed using the Welch-Satterthwaite approximation:

$$n_{df} = \frac{\left(\dfrac{s_1^2}{n_1} + \dfrac{s_1^2}{n_2}\right)^2}{\dfrac{s_1^4}{n_1^2(n_1 - 1)} + \dfrac{s_2^4}{n_2^2(n_2 - 1)}} = \frac{\left(\dfrac{v_1}{n_1} + \dfrac{v_1}{n_2}\right)^2}{\dfrac{v_1^2}{n_1^2(n_1 - 1)} + \dfrac{v_2^2}{n_2^2(n_2 - 1)}}$$

This formula may not produce an integer! (See Application Q31.)

---

23. What is the extra line of output in the Equal Variances table? What does its absence in the Unequal Variances output suggest?

---

t_{calc} is always **the difference in the means divided by the standard error in the difference.** When the variances are **not significantly different,** the estimate of the standard error can be improved by computing the pooled standard deviation. (The pooled variance listed in the output is the square of the pooled standard deviation.)

$$t_{calc} = \frac{|\bar{x}_1 - \bar{x}_2|}{s_{pooled}\sqrt{\dfrac{1}{n_1} + \dfrac{1}{n_2}}} \qquad \text{where} \quad s_{pooled} = \sqrt{\frac{(n_1 - 1)s_1^2 + (n_2 - 1)s_2^2}{n_1 + n_2 - 2}}$$

When the variances are **significantly different,** the pooled standard deviation is not computed, and the t_{calc} is computed using the original formula for t.

$$t_{calc} = \frac{|\bar{x}_1 - \bar{x}_2|}{\sqrt{\dfrac{s_1^2}{n_1} + \dfrac{s_2^2}{n_2}}}$$

24. Verify that these two different formulas for t_{calc} account for the different values of t Stat listed in the Excel output for Fred and Evan's data. Divide the work among your group.

25. Fill out the following table that contrasts the procedure for a t test based on the results of the F test:

	F test results	
	Variances Equal	**Variances Unequal**
Difference of means	$\|\bar{x}_1 - \bar{x}_2\|$	$\|\bar{x}_1 - \bar{x}_2\|$
Calculate s_{pooled}?		No
Standard error	$s_{pooled}\sqrt{\dfrac{1}{n_1} + \dfrac{1}{n_1}}$	
Formula for t_{calc}		$\dfrac{\|\bar{x}_1 - \bar{x}_2\|}{\sqrt{\dfrac{s_1^2}{n_1} + \dfrac{s_2^2}{n_1}}}$
Formula for n_{df}	$n_1 + n_2 - 2$	
Finding t_{crit}	Student's t table based on n_{df}	
Significant Difference?	YES if $t_{calc} > t_{crit}$	

26. What question does the *F* test answer?

27. Which of the three *t* tests does the *F* test affect, and how?

28. For group discussion: How confident are you that you understand when and how the *F* test and the *t* test are used?

29. On a scale from 1 to 10, rate your ability to read the statistical tables and Excel output.

Caution: The *F* test available in Excel's Data Analysis Toolpak **will compute *F* values less than one** if the first column of data entered into the test has the smaller variance. If you get output with an *F* value less than one, simply repeat the test but swap the positions of the data set ranges in the input window.

The *F* test is very simple to perform, so, as an alternative to the Toolpak *F* test, consider doing the test directly in the spreadsheet. The formula =VAR(range1)/VAR(range2) will give you the value of F_{calc}, and the formula =FINV(0.05, deg.fr(range1), deg.fr(range2)) will give you the value of F_{crit}. If your F_{calc} value is less than one, switch range1 and range2 in both formulas.

Applications

30. Using the F test in Excel, test the set of data you created for April in Q17. Show that April's and Betty's variances are significantly different.

31. Verify that the Welch-Satterthwaite formula yields the value 6.440 for Evan and Fred's data. To simplify the calculation, use the variances given in the Excel output.

32. Suppose you measured dissolved CO_2 in Brand X soda and Brand Y soda, and found that Brand X's soda had a standard deviation twice that of Brand Y. What is F_{calc}? Find a more complete F_{crit} table such as the one in D.C. Harris, *Quantitative Chemical Analysis* 8th Ed, page 81. Assume that you make the same number of measurements on both brands. Use the table to determine how many many measurements would you need in order to conclude that the difference in standard deviations is significant.

33. Add the value 12.9 to Evan's data and repeat the F test. Repeat <u>both</u> t tests as well. What happens to the t_{calc} values? Do the t tests give the same result?

34. Remove one of the 12.2 values from Fred's data and repeat the F test. Repeat <u>both</u> t tests as well. What happens to the t_{calc} values? Do the t tests give the same result?

35. Show that the two formulas for t_{calc} are equal when **either** a) $n_1=n_2$ **or** b) $s_1=s_2$.

36. Show that the Welch-Satterthwaite formula simplifies to $2n-2$ when $s_1=s_2$ and $n_1=n_2$.

37. Two students new to analytical chemistry each perform five identical acid-base titrations. They report standard deviations of 0.25 ml and 0.42 ml for the five titrations. Are their standard deviations significantly different?

38. The serum calcium concentrations of seven patients with hypoparathyroidism are compared with those of seven healthy patients. The seven patients with the disease showed calcium levels of 8.12, 7.98, 8.06, 8.20, 8.08, 8.15, and 8.22 mg/dL, while the seven healthy patients showed calcium levels of 8.82, 8.91, 9.23, 10.02, 9.67, 9.34, and 9.41 mg/dL. Is there a significant difference in the variance of serum calcium concentration for these two groups? Which t test must be used? Is there a significant difference in the average serum calcium concentration for these two groups?

39. Two new methods of measuring calcium in serum are compared. For five replicate measurements, the first method gave 9.23, 9.26, 9.23, 9.21, and 9.27 mg/dL, while the second method gave 9.20, 9.22, 9.23, 9.22, and 9.21 mg/dL. Do the two methods have significantly different means and standard deviations? Which t test must bc used?

Linear Regression for Calibration of Instruments

Learning Objectives

Students should be able to:

Content

- Describe linear regression as an averaging process.
- Interpret the results of a linear regression analysis.
- Evaluate the quality of a linear model for calibration of an instrumental measurement.

Process

- Construct and interpret graphs (Information processing)

Prior knowledge

- Mean, standard deviation.
- Instrumental calibration.
- Basic spreadsheet skills.

Further Reading

- Standard Quantitative Analysis Textbooks
- Anscombe's Quartet (classical model of problems with correlation coefficients as measures of fit)

  http://en.wikipedia.org/wiki/Anscombe%27s_quartet (accessed 8-17-13).

Author

Christopher F. Bauer

In a previous activity, you learned how to calibrate an instrument and evaluate figures of merit such as dynamic range and detection limit. We are revisiting calibration in order to explore the statistical procedures involved.

We'll begin by revisiting another idea with which you are already familiar—the mean or average—but with a new interpretation that will be developed through the Key Questions below.

Consider this...

A laboratory technician has been given a sample of household tap water in which to determine the concentration of lead. She removes six subsamples and determines the lead concentration in each using furnace atomic absorption spectrometry. (This instrumental measurement is based on the amount of light absorbed by atoms of lead produced by the furnace. The furnace is raised rapidly to a temperature of more than a thousand degrees C to decompose and vaporize the lead in a microliter-sized volume of sample.) The instrument signals and summary statistics are in Table 1.

Table 1 *Absorption measurements for lead in a single sample*

Trial	Signal in absorbance units
1	0.033
2	0.040
3	0.030
4	0.039
5	0.034
6	0.034

Mean	=	0.035
Standard Deviation	=	0.0038
Variance	=	1.44E-05
Sum of Squares	=	7.714E-05

Key Questions

1. Sketch a graph of signal magnitude vs trial number.

---

2. Draw a horizontal line on the graph at the value for the mean. Infer from the graph the values for the slope and intercept for this line.

3. As a group, decide how to describe the graph and its points in words, as if telling another person who cannot see the graph. Don't just repeat numerical values. Describe the position of points and the line. Write this down. Exchange your description with another group to check for understanding.

4. Calculate the differences between each point (y_i) and the mean ($\overline{y}$, called y bar). Draw lines on the graph to represent these differences.

5. Square each of these differences and add them together. Write an algebraic representation for this process. This is called the "sum of squared deviations from the mean" or simply the "sum of squares," sometimes abbreviated as SS.

6. Describe the ways in which the graph appearance would be different if the data set had a larger sum of squares, but the same mean. What would happen if it had a smaller sum of squares but the same mean?

7. Assume four other analysts approach you, each with a different suggestion about the appropriate central value for the data set: 0.030, 0.032, 0.038, or 0.040 absorbance units, respectively. Have each member of the group calculate the sum of squares for the data based on one of the four proposed central values.

8. Construct a graph of sum of squares vs. proposed value of mean. Include in the graph the original mean and sum of squares from Table 1 and the values calculated in Q7.

9. In your own words, describe the shape of this graph. Try to recall the mathematical term for the shape you see.

10. Consider the shape of the graph. What is unique about the correct mean (0.035 absorbance units) as an indicator of the central tendency of a set of measurements? Agree on and write down one or two concise sentences that summarize this idea.

Consider this...

Returning to the problem of measuring lead in household tap water by atomic absorption (AA) spectrometry, the laboratory technician calibrated the AA instrument by measuring the absorption signal for a set of standards containing different lead concentrations. The data are shown here in Table 2 and in Figure 1.

Table 2 Absorption calibration data for lead						
Lead Concentration (µg/L)	10.0	20.0	30.0	40.0	50.0	60.0
Instrument response (absorbance units)	0.015	0.043	0.054	0.085	0.101	0.122

Figure 1 Plotted points

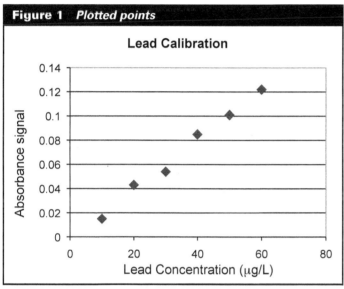

A larger version of this graph is at the end of the activity.

Key Questions

11. Theoretically, the atomic absorption signal should have a straight line relationship with the concentration of lead (Beer's Law). Does this data set seem to confirm that behavior?

12. If you were going to place a straight line on the graph, consider where it should be placed. Sketch a possible location for that line on your graph. Agree on and write down an explanation for why you decided to put it in that location.

13. Using Excel, it is possible to perform a *regression analysis,* which is a statistical procedure that determines the location of a function (e.g. in this case, a straight line) to best represent the position of the data set. Several types of output can be selected, in particular a Fitted Line Plot, a Residuals Table, and a Residuals Plot are shown below. Describe to each other what information you see here.

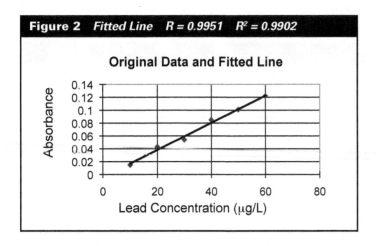

Figure 2 *Fitted Line* $R = 0.9951$ $R^2 = 0.9902$

Original Data and Fitted Line

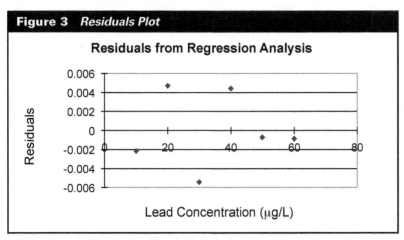

Figure 3 *Residuals Plot*

Residuals from Regression Analysis

Table 3 *Residuals Table*

Observation	Predicted Y	Residuals
1	0.017143	-0.002142857
2	0.038286	0.004714286
3	0.059429	-0.005428571
4	0.080571	0.004428571
5	0.101714	-0.000714286
6	0.122857	-0.000857143

14. How well does your rough line drawn in Q12 match the location of the "fitted line" in the plot shown in Figure 2?

15. In Table 3, look for the column "Predicted Y." Verify by inspection that these data define the predicted line in the graph Figure 2.

16. Inspect Figure 3 (called a *residuals plot*). Explain what the "Residuals" axis value of zero represents. In the Residuals Table 3, compare the column of "residuals" to the "residual plot." How are these related?

17. The Excel regression analysis determines the placement of the line by the "method of least squares." Remind yourselves of the discussion you had in Q1-Q10, and look at the Residuals Plot from the Excel analysis. What do you think "least squares" means in the context of fitting a line to data?

18. Explain what steps you would have take to generate a graph similar to that in Q7-Q8 to demonstrate that the regression line is a least squares result.

19. There is an intentional relationship between the graph you constructed in Q2-Q4 and the Residuals Plot discussed in Q16. Look at them carefully and comment. Based on this comparison and the value you calculated previously for the SS residuals in Q5, guess at the value for the SS Residuals for the regression analysis.

20. Verify this planned coincidence by calculating the variance from the SS Residuals (divide SS Residuals by (n-1), the number of degrees of freedom), and compare with the variance shown in Table 1. Comment.

Consider this...

Various indicators are used to express how closely the data fit a straight line function. Two common ones are the well-named Correlation Coefficient, R, and the poorly named Coefficient of Determination, R^2. The value of R^2 has the handy property of being equal to the fraction of *explained variance*—the amount of difference in the calibration measurements accounted for by fitting the straight line. Numerically:

$$R^2 = SS_{regression} / SS_{total}$$

SS_{total} is the sum of squared differences of each value from the mean of the values.

Key Questions

21. Each person in the group should take a turn at describing what R and R^2 mean in everyday language (that their parents or friends might be able to understand). Write those down.

22. What is the range of possible values of R and R^2? What values do you think represent a "good fit" in terms of calibration? After making this decision, review the values of R and R^2 for Figure 2 to see whether your intuition is on track.

Consider this...

In designing an instrumental measurement, an important consideration is establishing the reliability of the relationship of instrumental response to analyte quantity. Regression analysis determines what this relationship is. Regression analysis can also be used to monitor the stability of the relationship. Doing so is considered an essential part of quality control.

How will you know when something is wrong? Table 4 shows three sets of calibration data obtained for the same Atomic Absorption measurement of lead on different days. Linear regression analysis is conducted for each case. Below the table is Excel output: the values of R and R^2 and the Residuals Plot.

Table 4						
Lead Concentration (µg/L)	10.0	20.0	30.0	40.0	50.0	60.0
Case 1	0.011	0.041	0.059	0.077	0.092	0.100
Case 2	0.018	0.041	0.059	0.083	0.095	0.140
Case 3	0.014	0.016	0.034	0.036	0.054	0.056

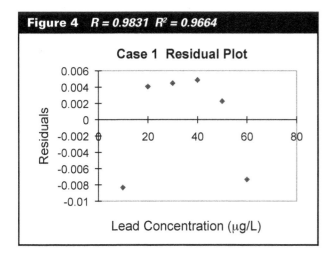

Figure 4 *R = 0.9831 R^2 = 0.9664*

Case 1 Residual Plot

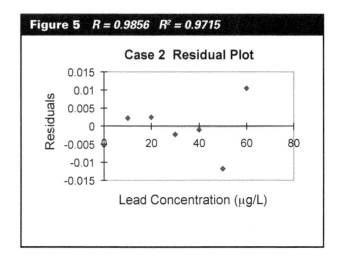

Figure 5 *R = 0.9856 R^2 = 0.9715*

Case 2 Residual Plot

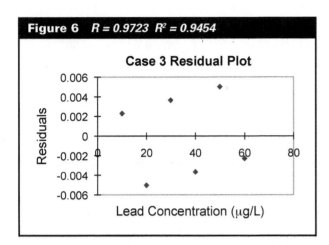

Figure 6 $R = 0.9723$ $R^2 = 0.9454$

Case 3 Residual Plot

Key Questions

23. Look at the residual plots (Figures 4-6). Each group member should write a sentence describing how the points are distributed on one of the plots. Then share these descriptions with each other. Be as specific and descriptive as you can. Come to an agreement on what the "problem" seems to be in each Case. Each Case demonstrates the failure of an underlying assumption in linear regression. Write your group consensus descriptions here.

24. Comment on what the values of R or R^2 tell you (or not) about the existence of potential problems in each case. It will help to refer back to Figures 2 and 3 as a case where these "problems" do not occur. You may wish to refer to the "sensitivity" of R and R^2 to these problems.

The statistical process of regression involves some important assumptions about the underlying errors (as indicated by the residuals). In particular, if the regression accounts for all of the systematic differences among the data points, then the residuals are just showing the random noise present in the method as it is being used. If there are unaccounted-for systematic differences at work among the data, it is important to know that, because this introduces error into measurements. It is important to look for evidence regarding these possible effects. In statistics terms, this is referred to as checking for *model fit, uniform variance, normality, and independence.*

Model fit	The instrumental response should be a linear function of the analyte concentration. ("Linear" is most desired for simplicity.)
Uniform variance	The size of errors should be about the same across the measurement range.
Normality	Errors should distribute in the shape of a normal curve.
Independence	Errors should not be correlated or related to each other.

25. Decide which problem is being shown by each of the three cases. Explain why you decided that.

Case   1   2   3

Problem   _____   _____   _____

One assumption is not included in Q25. It cannot be easily tested unless one has many data points. There are statistical tests and graphs designed to check this assumption, but the small data sets that are typical in many analytical situations do not often lead to definitive conclusions. You may have completed another activity about normal (Gaussian) distributions that delves more deeply into this issue.

26. As a group, consider whether each person has had an opportunity to contribute to the discussion and whether each person feels confident about his/her understanding of linear regression.

27. Each group member should identify one specific thing he/she learned about regression analysis.

28. The group should consider what aspect of regression analysis is still not clear to them. (Perhaps the application questions will help clarify.)

Applications

29. If you were the manager of the laboratory in which the three calibrations in Table 4 were generated, suggest a course of action that might correct the observed problems for these two different situations:

(a) All three cases were from a single analyst calibrating a single instrument on three different occasions.

(b) Each case was a different analyst working on different instruments.

30. Assume that you have a severe signal drift problem (magnitude of signal gradually changes in one direction over time). What problem would this introduce if you analyze your calibration standards in sequential order of lowest to highest concentration (a very common procedural approach)? Analyzing the standards in random sequence will eliminate this source of bias. Explain why. If you do not know whether drift problems exist, as lab manager, what would your recommendation be on sequencing standards?

31. It is true that if a set of data describes a straight line with non-zero slope, then the correlation coefficient will have a value close to 1. Is the converse necessarily true? If you find a set of data where the correlation coefficient is close to 1, does that mean the data describe a straight line? Take a position and write down your response. If you decide that the correlation coefficient does not necessary tell you "the data lie on a straight line," then what else can you do to make sure you don't come to the wrong conclusion?

32. What does a linear regression do and what meaning is attached to the residuals from a regression analysis? Write down an answer as an individual. Then, discuss until you achieve a group consensus on wording, and write that down.

33. Carry out the suggestion you put forward in Q17-Q18. Split up the work among the group members and combine your results.

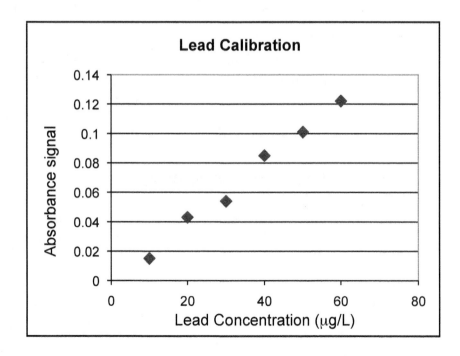

The Importance of Ionic Strength

Learning Objectives

Students should be able to:

Content

- Be able to describe the role of electrolytes in determining the ionic strength of solutions.

- Be able to explain, on a microscopic level, the effect of ionic strength on equilibrium concentrations.

- Be able to calculate the ionic strength of a solution and predict its impact on equilibrium concentrations.

Process

- Identifying key properties of electrolyte solutions. (Problem solving)

- Depicting a shared understanding of microscopic solution interactions. (Information processing, teamwork)

- Challenging assumptions. (Critical thinking)

Prior knowledge

- Be able to calculate equilibrium concentrations for solution species.

- Be familiar with precipitation reactions and the solubility of sparingly-soluble salts.

- Be familiar with the basic properties of an electrolyte.

- Be familiar with Le Châtelier's Principle and the impact of common ions on equilibrium compositions.

Authors

Juliette Lantz, David Langhus and Shirley Fischer-Drowos

Consider this...

Suppose that we are given the following reaction and the accompanying solubility product equilibrium constant:

$$AgBr(s) \rightleftharpoons Ag^+(aq) + Br^-(aq)$$

$$K_{sp} = 5.0 \times 10^{-13}$$

Solubility is defined as the amount of solid that dissolves in a solution, usually expressed in units of mol/L or g/L.

Key Questions

1. Calculate the $[Ag^+]$ and $[Br^-]$ in a saturated solution of AgBr(s) in deionized water.

2. Consider the equilibrium of AgBr(s) in an aqueous solution. According to Le Châtelier's Principle,

    a)   **adding** some additional Ag^+ ions would cause $[Br^-]$ to

                 INCREASE   DECREASE   NOT CHANGE

                 Briefly provide your rationale:

    b)   **removing** some Ag^+ ions would cause $[Br^-]$ to

                 INCREASE   DECREASE   NOT CHANGE

                 Briefly provide your rationale:

    c)   **adding** some KNO_3 would cause $[Br^-]$ to

                 INCREASE   DECREASE   NOT CHANGE

               Briefly provide your rationale:

Consider this...

KNO_3 is a strong electrolyte, that is, it completely dissociates in water to yield a stoichiometric quantity of ions. It has been added in varying concentrations (as shown in Table 1) to a saturated solution of AgBr(s).

Table 1		
Solution	Initial electrolyte concentration	$[Ag^+] = [Br^-]$
Deionized water	0	7.1×10^{-7} M
KNO_3	1.0×10^{-5} M	7.1×10^{-7} M
	1.0×10^{-4} M	7.2×10^{-7} M
	1.0×10^{-3} M	7.3×10^{-7} M
	1.0×10^{-2} M	7.9×10^{-7} M
	1.0×10^{-1} M	9.5×10^{-7} M

Key Questions

3. Does the value you calculated in Q1 agree with any of those shown in Table 1? If so, which ones?

4. Examine the data in Table 1. As $[KNO_3]$ increases, do $[Ag^+]$ and $[Br^-]$

        INCREASE   DECREASE   STAY THE SAME

5. Is your answer to Q4 consistent with your prediction in Q2c?

6. In a grammatically correct sentence, summarize the impact of increasing the KNO_3 concentration on the solubility of AgBr(s).

> **Ionic strength** is a measure of the total ion concentration of a solution. It is given the symbol μ and has units of molarity.

7. As a group, brainstorm the properties of an ion that will affect its ability to contribute to the ionic strength of the overall solution.

> Consider the Ag^+ ion. Since it has a positive charge it will tend to attract any negatively charged ions present in solution. When it appears in a solution of distilled water in which there aren't many other ions available, you can imagine that it is surrounded by polar water molecules, with the negative end of several water molecules in close proximity to the positively charged ion.
>
> When the ion is in a solution containing other ions, some of the oppositely charged ions are attracted to it, just like one end of the polar water molecule. The ions that closely surround the equilibrium ion are referred to as the "ionic atmosphere" for that ion. The whole assembly is called the "solvated ion."

8. When Ag^+ (from solid AgBr) appears in an aqueous solution of KNO_3, which ions will be in close proximity to it?

9. When Br^- (from solid AgBr) appears in a solution of KNO_3, which ions will be in close proximity to it?

10. As a group, draw a picture below showing Ag^+ surrounded by the ions that comprise its ionic atmosphere. Draw another picture below showing Br^- surrounded by its ionic atmosphere. In your pictures, represent the ions in relative amounts, and be sure to also include some water molecules. When you are done, compare your pictures with those produced by other teams.

11. Does the ionic atmosphere around the silver ion enhance or diminish its +1 charge? Justify your choice.

12. Does the ionic atmosphere around the bromide ion enhance or diminish its -1 charge? Justify your choice.

13. Do the ionic atmospheres surrounding Ag^+ and Br^- enhance or diminish their electrostatic attraction for each other? Explain your reasoning.

14. If the charges on the Ag^+ and Br^- ions are surrounded by their respective ionic atmospheres, do you expect the solubility of AgBr(s) to

INCREASE   DECREASE   STAY THE SAME

What information did you use to determine this? Compare your answers as a group, and provide information that supports your response. Verify your answer by looking at Table 1. What happens to the [Ag^+] and [Br^-] as the concentration of the inert salts increases?

15. In a grammatically correct sentence, explain why increasing electrolyte concentration has the effect on the solubility of AgBr(s) that is observed.

Consider this...

$Mg(NO_3)_2$ is another strong electrolyte which dissociates completely in water. It has been added in varying concentrations (as shown in Table 2) to a saturated solution of AgBr(s).

Table 2		
Solution	Initial electrolyte concentration	$[Ag^+] = [Br^-]$
Deionized water	0	7.1×10^{-7} M
KNO_3	1.0×10^{-5} M	7.1×10^{-7} M
	1.0×10^{-4} M	7.2×10^{-7} M
	1.0×10^{-3} M	7.3×10^{-7} M
	1.0×10^{-2} M	7.9×10^{-7} M
	1.0×10^{-1} M	9.5×10^{-7} M
$Mg(NO_3)_2$	1.0×10^{-5} M	7.1×10^{-7} M
	1.0×10^{-4} M	7.2×10^{-7} M
	1.0×10^{-3} M	7.6×10^{-7} M
	1.0×10^{-2} M	8.7×10^{-7} M
	1.0×10^{-1} M	1.2×10^{-6} M

Key Questions

16. Examine the data in Table 2. As the initial concentration of $Mg(NO_3)_2$ increases, do $[Ag^+]$ and $[Br^-]$

                INCREASE   DECREASE   STAY THE SAME

17. Now, look at the solubility of AgBr(s) at the same concentration (perhaps 1×10^{-3} M) of the two different electrolytes (KNO_3 and $Mg(NO_3)_2$) in Table 2. With which electrolyte is AgBr(s) more soluble?

18. As a group, list the differences in the properties of solutions of a solution of $Mg(NO_3)_2$ and a solution of KNO_3 (with equal concentrations) that might cause them to have different impacts as electrolytes.

19. Assume $(NH_4)_2X$ (s) was a soluble ionic salt used to make up an electrolyte solution. At similar concentrations, would you expect it to behave more similarly to a solution of $Mg(NO_3)_2$ or a solution of KNO_3? Justify your choice.

20. Why are KNO_3 and $Mg(NO_3)_2$ considered to be strong electrolytes, while a saturated solution of AgBr(s) isn't?

Consider this...

The ionic strength of a solution (μ) is:

$$\mu = \frac{1}{2}\sum_{ions\ i} C_i z_i^2$$

where z is the charge on an ion and C is its concentration in units of molarity.

The total number of ions in a solution changes the behavior of the ions in equilibrium (in this case $[Ag^+]$ and $[Br^-]$.)

21. Consider an electrolyte solution of 1.0×10^{-2} M KNO_3. Calculate the following and verify your answers with your group:

a) The total concentration of all the ions in this electrolyte solution.

b) The ionic strength of this solution.

22. Consider an electrolyte solution of 1.0×10^{-2} M $Mg(NO_3)_2$. Calculate the following and verify your answers with your group.

a) The total concentration of all the ions in this electrolyte solution.

b) The ionic strength of this solution.

23. a) Fill in Table 3 below with your answers from Q21 and Q22.

b) Support or refute this claim using evidence from Table 3, or refine it as needed:
"A slightly soluble salt such as AgBr(s) will become more soluble as the total ion concentration of the solution increases."

Table 3				
Solution	**Initial electrolyte concentration**	**Total ion concentration**	**Ionic strength (μ)**	**$[Ag^+] = [Br^-]$**
Deionized water	0	0	0	7.1×10^{-7} M
KNO_3	1.0×10^{-2} M			7.9×10^{-7} M
KNO_3	1.5×10^{-2} M	3.0×10^{-2} M	1.5×10^{-2} M	8.0×10^{-7} M
$Mg(NO_3)_2$	1.0×10^{-2} M			8.7×10^{-7} M

24. As a group, generate a list of the new properties you learned about solutions from this activity.

25. How did your team reach consensus regarding the picture representing the microscopic solution interactions?

Applications

Note – Application questions 26-30 refer to the slightly soluble AgBr(s) system described in this activity, with varying concentrations of the supporting electrolyte.

26. Calculate the ionic strength (μ) of a) 1.0×10^{-3} M KNO_3 and b) 1.0×10^{-3} M $Mg(NO_3)_2$. Verify that the ionic strength of $Mg(NO_3)_2$ is indeed greater than that of KNO_3 at the same concentration.

27. Do the concentrations of Ag^+ and Br^- resulting from the dissolution of AgBr(s) contribute significantly to the ionic strength of either of these solutions? Explain.

28. Calculate the ionic strength of 0.10 M $NaNO_3$. What do you expect the concentration of Ag^+ and Br^- to be in this electrolyte?

29. Calculate the ionic strength of 0.010 M Na_2SO_4. What do you expect the concentration of Ag^+ and Br^- to be in this electrolyte?

30. Calculate the ionic strength of 0.020 M KBr plus 0.010 M Na_2SO_4. Do you expect the concentrations of Ag^+ and Br- to be higher or lower than those in Application Question 29?

31. What is the ionic strength of a solution containing 0.010 M $MgSO_4$ and 0.050 M HCl?

32. An electrode was constructed in lab that measures equilibrium silver ion concentrations in solutions saturated with particular insoluble silver salts such as AgCl(s). Effective measurement of the silver ion at these trace levels requires that the ionic strength of the solutions in which Ag^+ is measured be held constant.

    a) In Step 1 of the experiment, 10.0 mL of a solution containing 0.0010 M $AgNO_3$ and 0.100 M NH_4NO_3 is diluted to 100.00 mL in a volumetric flask. What is the resulting ionic strength of this solution?

    b) To the 100.00 mL solution made in Step 1, 10.00 mL of a solution containing 0.0100 M NH_4Cl is added. Verify that the ionic strength has in fact been kept constant, assuming that stoichiometric amounts of AgCl are formed, with Ag^+ acting as the limiting reagent.

33. This figure is used with permission from the following article: Marzzacco, C.J. 1998. "The Effects of Salts and Nonelectrolytes on the Solubility of Potassium Bitartrate." J. Chem. Ed. 75 (12): 1628.

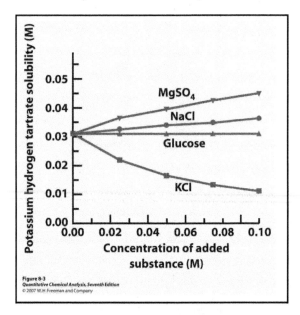

Figure 8-3
Quantitative Chemical Analysis, Seventh Edition
© 2007 W.H.Freeman and Company

Explain the trends shown in this graph.

a) Why doesn't the solubility of potassium hydrogen tartrate appear to change upon increasing additions of glucose?

b) Why does the solubility of potassium hydrogen tartrate increase as NaCl is added?

c) Why does the solubility of potassium hydrogen tartrate show a greater increase with the addition of $MgSO_4$ than with the addition of NaCl?

d) Why does KCl decrease the solubility of potassium hydrogen tartrate unlike NaCl? Why is the effect more dramatic than that seen with NaCl?

34. It's desirable for systems containing water such as boilers, holding pools, plumbing and such to be protected from corrosion on the inside by a thin coating of water-insoluble $CaCO_3$. Unfortunately, in the real world, a few things can happen that might interfere with this. If the water is soft (low or zero Ca^{+2} concentration), this thin layer of $CaCO_3$ can dissolve over time, leaving the surfaces prone to corrosion. Such water is referred to in this context as aggressive. Conversely, if the water is very hard (high Ca^{+2} concentration), excessive amounts $CaCO_3$ can form. This is a particular problem when the water is heated. Such water is referred to as scaling. Water that is neither aggressive nor scaling is said to be in balance.

As you might imagine, ionic strength plays a role in the matter of balance. Suppose that one has the choice of filling a large industrial boiler system with rainwater (for all intents and purposes distilled) or processed sewer water (from which the gross stuff has been removed but not all the sodium ions, chloride ions, sulfate ions, etc. that found their way in there during the journey from the tap). For which of rain water and reclaimed sewer water will $[Ca^{+2}]$ have to be higher to achieve balance? Explain how you arrived at your decision.

Activity and Activity Coefficients

Learning Objectives

Students should be able to:

Content

- Describe the impact of ionic strength on activity coefficients and the equilibrium concentrations of slightly soluble salts.

- Develop strategies for calculating activity coefficients, activities and equilibrium concentrations in various electrolyte solutions.

- Discern the ionic strength range for which activities need to be used in equilibrium calculations.

Process

- Finding, predicting trends in data. (Information processing)

- Identifying similarities and differences in solutions. (Critical thinking)

- Estimating values, choosing solving strategies for various concentration regions. (Problem solving)

Prior knowledge

- Be familiar with the basic properties of an electrolyte.

- Be able to calculate the ionic strength of an electrolyte solution.

- Be familiar with precipitation reactions and the solubility of sparingly-soluble compounds.

- Be able to calculate equilibrium concentrations for solution species using equilibrium constants

Further Reading

- Harris, D.C. 2007. *Quantitative Chemical Analysis,* 7th Edition, Section 8-2, p.143. New York: W.H. Freeman. Section 15H, p.419. Belmont, CA: Thomson Brooks/Cole.

Authors

Juliette Lantz and David L. Langhus

Consider this...

$$AgBr(s) \rightleftharpoons Ag^+(aq) + Br^-(aq)$$

$$K_{sp} = 5.0 \times 10^{-13}$$

Table 1

Solution	electrolyte conc.	μ (Ionic strength) **	γ_{Ag^+} (Activity coefficient for Ag⁺)	γ_{Br^-} (Activity coefficient for Ag⁺)	$[Ag^+] = [Br^-]$
Distilled water	0		1	1	7.07×10^{-7} M
KNO$_3$	1×10^{-5} M	1×10^{-5} M	0.996	0.996	7.1×10^{-7} M
	1×10^{-4} M	_____	0.988	0.988	7.2×10^{-7} M
	1×10^{-3} M	1×10^{-3} M	0.964	0.964	7.3×10^{-7} M
	1×10^{-2} M	_____	0.897	0.897	7.9×10^{-7} M
	1×10^{-1} M	1×10^{-1} M	0.745	0.745	9.5×10^{-7} M
Mg(NO$_3$)$_2$	1×10^{-5} M	3×10^{-5} M	0.994	0.994	7.1×10^{-7} M
	1×10^{-4} M	_____	0.980	0.980	7.2×10^{-7} M
	1×10^{-3} M	3×10^{-3} M	0.940	0.941	7.5×10^{-7} M
	1×10^{-2} M	_____	0.837	0.840	8.4×10^{-7} M
	1×10^{-1} M	3×10^{-1} M	0.642	0.658	1.1×10^{-6} M

**For these calculations, it is assumed that these soluble compounds fully dissociate into cations and anions. The possibility of ion pairs is discussed in Application 22.

Key Questions

1a. In Table 1, are the concentrations of the two electrolyte series (KNO$_3$ and Mg(NO$_3$)$_2$) the same?

1b. In Table 1, are the ionic strengths of the two electrolyte series (KNO$_3$ and Mg(NO$_3$)$_2$) the same?

2. Divide the following tasks among your group. Using the formula for ionic strength

$$\mu = \frac{1}{2} \sum_{ions} C_i z_i^2$$

calculate μ for these separate solutions:

a. 1.0×10^{-2} M KNO_3 and 1.0×10^{-2} M $Mg(NO_3)_2$

b. 1.0×10^{-4} M KNO_3 and 1.0×10^{-4} M $Mg(NO_3)_2$

Use your calculations to verify that the ionic strength of $Mg(NO_3)_2$ is greater than that of KNO_3 at the same concentration.

3. Add your calculated values to Table 1 in the appropriate spaces. Examine Table 1, and decide if your answers to Q2 are correct.

Consider this…

A saturated solution of AgBr *(s)* in deionized water contains tiny concentrations of Ag^+ and Br^-, and in fact we can actually calculate what the concentrations of these ions are using the K_{sp}. When an electrolyte such as KNO_3 is added to such a saturated solution, the ionic atmospheres of Ag^+ and Br^- accumulate some electrolyte ions in proportion to how many of the electrolyte ions are present. This causes the charges on Ag^+ and Br^-, which are largely responsible for how these ions interact in solution, to appear somewhat diminished. This diminishing of the charges on the Ag^+ and Br^- ions by their ionic atmospheres could be thought of as diminishing their "effective concentrations."

The **activity** of an ion, or its "effective concentration" due to the presence of an electrolyte, is given as:

$$\text{Activity} = A_i = [i]\gamma_i$$

where A_i is the activity of ion i, [i] is its molar concentration, and γ_i is the so-called activity coefficient, a proportionality constant between molar concentration and activity that quantifies how much the effective concentration of ion i has been diminished. Note that γ values are dependent upon the particular ion in question and the ionic strength of a solution.

4. Examine the activity coefficients for Ag^+ in Table 1. What happens to the magnitude of the activity coefficient, γ_{Ag^+}, as the concentration of KNO_3 increases?

5. What happens to the magnitude of the activity coefficient, γ_{Br^-}, as the concentration of the KNO_3 increases?

6. Compare the changes in the magnitudes of γ_{Ag^+} and γ_{Br^-} in KNO_3 to those in $Mg(NO_3)_2$. Which electrolyte is having a greater impact on these activity coefficients?

7. In **low ionic strength** solutions –

    ➡ μ becomes small, as shown in Table 1. What value will γ approach?

    ➡ Under these conditions would activity (A_i) be

       (less than)   (greater than)   (about equal to)   concentration?

       Support your answer:

8. In **high ionic strength** solutions –

    ➡ μ becomes large, according to Table 1. What happens to γ?

    ➡ Under these conditions would activity (Ai) be

       (less than)   (greater than)   (about equal to)   concentration?

       Support your answer:

9. Support or refute this statement, providing justification for your position that includes a sketch showing the atmosphere surrounding a neutral species in an aqueous solution:

"For a neutral species in an electrolyte solution, it is reasonable to expect that γ is approximately one."

10. In a complete, grammatically correct sentence, summarize the dependence of activity coefficients on the concentration of an electrolyte in solution.

11. Consider Table 1 again. As the concentration of either electrolyte (KNO_3 or $Mg(NO_3)_2$) increases, what happens to $[Ag^+]$ and $[Br^-]$? For which electrolyte is there a more dramatic change?

12. In a complete, grammatically correct sentence, summarize the effect of ionic strength on the solubility of a sparingly soluble salt such as AgBr(s).

Consider this...

By now, you've no doubt noticed that the method you learned in an earlier course for calculating [Ag⁺] and [Br⁻] in a saturated solution of AgBr*(s)* works well for solutions having ionic strengths that are close to zero, but not as well when the ionic strength starts to get higher. This is because K_{sp} is actually an "activity constant." It's only really a constant if you use chemical activities in your calculation rather than molar concentrations.

To solve an equilibrium problem using activities, an activity coefficient, γ, is needed for **each** ion in solution; the value of each γ is dependent upon the ionic strength (μ) of the solution. Recall that the ionic strength (μ) of a solution can be calculated by taking into account the concentration of all ions present in appreciable amounts and their corresponding charges. There are two ways to obtain an activity coefficient, γ, for any ion in solution:

Method 1. Tables exist that list activity coefficient values for many common ions in solutions of varying ionic strength. One is likely to be found in your textbook.

Method 2. Activity coefficients can be calculated using the Extended Debye-Huckel (EDH) equation, in low ionic strength conditions $(\mu < 0.1 \text{ M})$.

$$\log \gamma = \frac{-0.51z^2 \sqrt{\mu}}{1 + \frac{\alpha \sqrt{\mu}}{305}}$$

where μ is the ionic strength of the solution, z is the charge on the ion of interest, and α is the hydrated ionic radius in pm, which is also likely available in a table in your textbook.

13. Divide up the following tasks among your group; report back to the whole group and verify each conclusion:

    a. For a solution of 1 x 10⁻³ M KNO₃, calculate γ for Ag⁺ using the EDH equation. Does it agree with the γ value found in Table 1?

    b. For a solution of 1 x 10⁻³ M KNO₃, calculate γ for Br⁻ using the EDH equation. Does it agree with the γ value found in Table 1?

    c. For a solution of 1 x 10⁻³ M Mg(NO₃)₂, calculate γ for Ag⁺ using the EDH equation. Does it agree with the γ value found in Table 1?

    d. For a solution of 1 x 10⁻³ M Mg(NO₃)₂, calculate γ for Br⁻ using the EDH equation. Does it agree with the γ value found in Table 1?

Consider this...

The equilibrium constant values tabulated in your textbook and most other places are "activity constants." That is, they're strictly applicable only if you use chemical activities for the concentration terms. So, for example, we should correctly write the solubility product expression for AgBr(s) as

$$K_{sp} = 5.0 \times 10^{-13} = A_{Ag+} \times A_{Br^-}$$

where A_{Ag+} represents the activity of Ag^+ and A_{Br^-} represents the activity of Br^-.

If the activities of Ag^+ and Br^- are used to correctly represent the behavior of these ions in a solution where many ions are present, the expression becomes

$$K_{sp} = A_{Ag+} \times A_{Br^-} = [Ag^+]\gamma_{Ag+}[Br^-]\gamma_{Br^-}$$

14. Now, using the new equilibrium expression (and dividing the following tasks up among your group members)

$$K_{sp} = [Ag^+]\gamma_{Ag+}[Br^-]\gamma_{Br^-}$$

a. Calculate $[Ag^+]$ and $[Br^-]$ in a 1×10^{-3} M solution of KNO_3. Compare your results to those found in Table 1.

b. Calculate $[Ag^+]$ and $[Br^-]$ in a 1×10^{-3} M solution of $Mg(NO_3)_2$. Compare your results to those found in Table 1.

15. Consider the equilibrium involving the auto-ionization (hydrolysis) of water:

$$H_2O(l) \rightleftharpoons H^+(aq) + OH^-(aq) \qquad K_w = 1.00 \times 10^{-14}$$

a. Write the K_w expression corresponding to this equilibrium as you would have in an earlier course, using [H⁺] to represent the molar concentration of $H^+(aq)$ and [OH⁻] to represent the molar concentration of $OH^-(aq)$.

b. Now, write the expression for K_w using the activities of $H^+(aq)$ and $OH^-(aq)$, A_{H^+} and A_{OH^-}.

c. Modify the expression that you wrote for K_w in Q15b) to include activity coefficients and concentrations of $H^+(aq)$ and $OH^-(aq)$.

d. So suppose we have this 0.0100 M solution of KNO_3 in water. KNO_3 doesn't contain any H^+ or OH^- so the only source of these ions is the autoionization of water. Every H^+ we find in solution means that there's exactly one OH^- floating around too, so [H⁺] = [OH⁻]. Find values for γ_{H^+} and γ_{OH^-} and do the necessary math to solve for [H⁺].

e. pH is defined in terms of the activity of H^+. pH sensors respond to activity, so when you use a pH meter it's the activity of H^+ you're observing, not [H⁺]. pH can be calculated from [H⁺] though using:

$$pH = -\log_{10}([H^+]\gamma_{H^+})$$

Calculate the pH of that 0.0100 M KNO_3 solution in Q15d.

16. As a group, decide whether you want to support or refute the following statement and provide your justification.

"The use of activities is required to describe equilibrium concentrations when initial electrolyte concentrations equal or exceed 1×10^{-3} M."

17. Brainstorm together. Can you think of any laboratory measurements where adding an electrolyte to a solution might be necessary?

18. In what ways have the equilibrium problem solving strategies you learned in general chemistry been challenged or expanded?

19. How did your group ascertain when a calculation was performed correctly? What strategies did you use to ensure that all group members could perform the calculations correctly?

Applications

20. A saturated solution of AgI contains 0.05 M KNO_3. The K_{sp} for AgI is 8.3×10^{-17}.

    a. As a first approximation, assume that the ionic strength of this solution is dominated by 0.05 M KNO_3. Calculate μ.

    b. Using the EDH, calculate γ_{Ag+} and γ_{I-}.

    c. Taking activities into account, calculate $[Ag^+]$ and $[I^-]$ in this solution.

    d. Was the approximation in Q20a valid? It not, how could you amend your ionic strength calculation?

21. In this activity we considered KNO_3 solutions in the range 1.0×10^{-1} to 1.0×10^{-5} M. The Extended Debye-Hückel equation is a reasonable way to compute activity coefficients for $\mu \leq 0.1$ M. Suppose we have AgBr(s) dissolved in a 1.0 M solution of KNO_3. At this ionic strength will the activity coefficients be closer to or further from 1 than they are for the lower concentration KNO_3 solutions?

22. Magnesium sulfate is a soluble salt. Suppose you prepare a 0.025 M solution of $MgSO_4$.

    a. Write the dissolution reaction that occurs when $MgSO_4(s)$ dissolves in water.

    b. In reality, the following ion pairing reaction also occurs in solutions of $MgSO_4$:

$$Mg^{2+}(aq) + SO_4^{2-}(aq) \rightleftharpoons MgSO_4(aq)$$

The ion-pair formation constant K is the equilibrium constant for this reaction and has the form

$$K = \frac{[MgSo_4]\gamma_{MgSO_4}}{[Mg^{2+}]\gamma_{Mg^{2+}}[SO_4]\gamma_{SO_4^{2-}}} = 170$$

Use the systematic treatment of equilibrium (hint: write two mass balance equations) to find $[Mg^{2+}]$ and $[SO_4^{2-}]$. Disregard any other chemical reactions and activity coefficients.

Now we will consider the ionic strength and activity effects on this equilibrium.

    c. Use your answer from Q22b to calculate the ionic strength of this solution and the activity coefficients for Mg^{2+} and SO_4^{2-}.

    d. Solve the systematic equilibrium problem posed in Q22b again taking activity coefficients into account.

    e. Repeat the calculations of Q22c and Q22d two more times in order to find your best estimate of the Mg^{2+} concentration.

    f. The percent of $MgSO_4$ that is ion paired in solution is determined by the ratio of the concentration of $MgSO_4(aq)$ to the formal concentration of magnesium ($[Mg^{2+}]$ and $[MgSO_4]$). What percentage of the $MgSO_4$ is ion paired?

23. Consider two solutions of weak acids, as described below.

Solution A	Solution B
0.010 M picric acid (pK_a = 0.38)	0.010 M acetic acid (pK_a = 4.75)

a. For each monoprotic weak acid, write the acid dissociation reaction that occurs in aqueous solution.

b. Disregarding activities, calculate the pH of each solution.

Now we will consider activity effects for these two acids.

c. Calculate the ionic strength of each solution. Note that the only ions in solution are those originating from the acid dissociation.

d. In which solutions will the activity coefficients deviate substantially from 1? In which solutions are the number of ions in solution sufficiently low so that $\mu \cong 0$?

e. If the activity coefficients deviate from 1, recalculate the pH of the solution taking activity into account. Iterate your calculations until you get the same pH to two decimal places on subsequent iterations.

24. Barium enema is a diagnostic medical procedure in which the inside of the large intestine is coated with an aqueous slurry of insoluble $BaSO_4$ and imaged using X-ray. If you're a chemist, this probably rings all sorts of alarm bells because Ba is a heavy metal that's absorbed throughout the gastrointestinal tract and levels in drinking water as low as 10 mg/L (about 7×10^{-5} M) have been shown to result in an elevated risk of heart attack and stroke. If you ask the medical folks they'll say "Oh, that's not a problem because $BaSO_4$ is completely insoluble in water." Of course you know that's questionable because your Quantitative Chemistry textbook lists a solubility product for $BaSO_4$ so there's got to be a little Ba^{2+} in that solution.

a. So what is the molar concentration of Ba^{2+} in a saturated solution of $BaSO_4$ in deionized water?

b. Of course they don't make this slurry in deionized water. They make it up in a saline solution that contains 0.137 M NaCl, 0.0027 M KCl, 0.0809 M Na_2HPO_4 (which dissociates into Na^+ and HPO_4^{2-} ions) and 0.0176 M KH_2PO_4 (which dissociates into K^+ and $H_2PO_4^-$ ions). Assuming that no acid-base shenanigans take place, what's the ionic strength of this solution, ignoring the contributions of Ba^{2+} and SO_4^{2-} that'll be present?

c. What's the molar concentration of Ba^{2+} in that slurry *really?*

d. In calculating the ionic strength in Q24b we assumed that the Ba^{2+} and the SO_4^{2-} didn't contribute significantly. Is that a reasonable assumption?

Multiple Equilibria: When Reactions Compete

Learning Objectives

Students should be able to:

Content

- Explain how a competing equilibrium affects the solubility of a sparingly soluble compound in water.

- Predict the pH dependence of the solubility of a sparingly soluble salt.

Process

- Identifying key reactions. (Problem solving)

- Challenging assumptions. (Critical thinking)

- Predicting pH dependence. (Information processing)

Prior knowledge

- Be familiar with the weak acid-base properties of ions at the General Chemistry level.

- Understand the principle of chemical equilibrium.

- Be able to calculate the concentration of ions in a saturated solution of a sparingly soluble compound at the General Chemistry level.

- Be familiar with the Le Châtelier's principle.

Further Reading

- Harris, D.C. 2007. Quantitative Chemical Analysis, 7th Edition, Section 8-4, 8-5, p.147; 13-1 – 13-3, p. 251. New York: WH Freeman.

- Skoog, D.A., D.M. West, F.J. Holler, and S.R. Crouch. 2004. Fundamentals of Analytical Chemistry, 8th Edition, Chapter 11, p. 281. Belmont, CA: Thomson Brooks/Cole.

Authors

David L. Langhus and Juliette Lantz

PGLANA004016a

Consider this...

$$AgCN(s) \rightleftharpoons Ag^+ + CN^- \qquad K_{sp} = 2.19 \times 10^{-16}$$

A series of saturated solutions of AgCN in water buffered at several pH values are prepared. The buffers are selected so that they contain neither Ag^+ nor CN^- and don't react with either of these ions. Additionally, the buffer solutions are prepared in such a way that all have a matching ionic strength of 0.1 M. The solutions are allowed to stand overnight with slow stirring to be sure that equilibrium is attained.

The solutions are allowed to settle and a sample of clear supernatant liquid is withdrawn from each. The molar concentration of silver ion in each of these saturated solutions was then determined:

Table 1 *Some species concentrations in saturated solutions of AgCN at various pH values.*

pH	[Ag⁺], M	[CN⁻], M	[HCN], M
1.00	1.88×10^{-4}	1.17×10^{-12}	1.88×10^{-4}
2.00	5.96×10^{-5}	3.69×10^{-12}	5.96×10^{-5}
3.00	1.88×10^{-5}	1.17×10^{-11}	1.88×10^{-5}
4.00	5.96×10^{-6}	3.69×10^{-11}	5.96×10^{-6}
5.00	1.88×10^{-6}	1.17×10^{-10}	1.88×10^{-6}
6.00	5.96×10^{-7}	3.69×10^{-10}	5.95×10^{-7}
7.00	1.89×10^{-7}	1.16×10^{-9}	1.88×10^{-7}
8.00	6.14×10^{-8}	3.58×10^{-9}	5.78×10^{-8}
9.00	2.40×10^{-8}	9.18×10^{-9}	1.48×10^{-8}
10.00	1.60×10^{-8}	1.38×10^{-8}	2.22×10^{-9}
11.00	1.50×10^{-8}	1.47×10^{-8}	2.37×10^{-10}
12.00	1.48×10^{-8}	1.48×10^{-8}	2.39×10^{-11}
13.00	1.48×10^{-8}	1.48×10^{-8}	2.39×10^{-12}

Key Questions

1. From your experience with chemistry to this point you're no doubt aware that some solids, like NaCl and $CuSO_4$, are quite soluble in water while others, like AgCl and CuS, are fairly insoluble. Examine the value of the K_{sp} for AgCN given above. Decide as a group if AgCN should be described as quite soluble in water or fairly insoluble.

2. Now look at Table 1. Does AgCN actually appear quite soluble in water or fairly insoluble? How did you decide?

You'll recall from your introductory knowledge of equilibrium that one can actually calculate the [Ag$^+$] one would expect to observe in a saturated solution of AgCN given its K_{sp}. At equilibrium a tiny amount of AgCN dissolves, giving small concentrations of Ag$^+$ and CN$^-$. This dependence is governed by the equilibrium expression:

$$K_{sp} = 2.19 \times 10^{-16} = [Ag^+][CN^-]$$

Key Questions

3. When AgCN*(s)* dissolves in solutions like these, what would you expect the relationship between [Ag$^+$] and [CN$^-$] to be?

4. Calculate [Ag$^+$] and [CN$^-$] in a saturated solution of AgCN in water. Agree among the members of your group on what these values are.

5. Compare the [Ag$^+$] you got above with the values in Table 1. Over what pH range does there appear to be good agreement between your calculated results for [Ag$^+$] and the actual values shown in Table 1?

6. Is there any pH range over which your calculation does not appear to give reliable results according to Table 1?

7. You probably assumed that $[Ag^+] = [CN^-]$ in order to do your calculation. Over what pH range does this appear to be a good assumption?

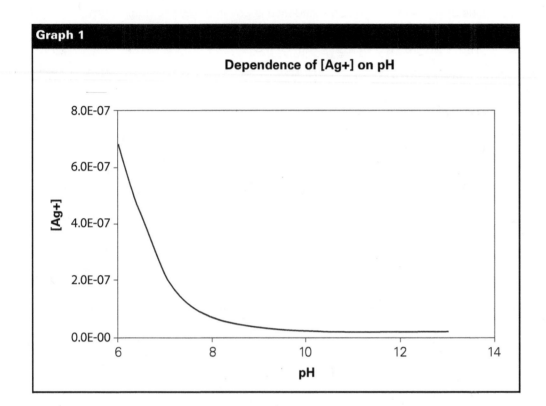

Graph 1

Dependence of [Ag+] on pH

y-axis: [Ag+]

x-axis: pH

Key Questions

Here's another way of looking at some of the data in Table 1:

8. Referring to Table 1 or Graph 1, as pH increases (the solution becomes more basic) what happens to the concentration of Ag^+?

9. Where does this Ag^+ come from?

10. Based on your answers to Q8 and Q9, what appears to happen to the solubility of AgCN as the pH increases?

11. Of the solutions shown in Table 1, what is the pH of that solution in which AgCN is most soluble?

12. You'll notice that Table 1 includes a column showing [HCN] in each solution. What happens to the concentration of HCN as the solution becomes more acidic?

13. When preparing the solutions for Table 1, buffers were used that contained no Ag^+ and no CN^-. Yet particularly at acidic pH, lots of HCN is present. As a group formulate an explanation for where this HCN is coming from.

14. CN^- is known to be a weak base in water for which $K_b = 1.6 \times 10^{-5}$, and you might remember from General Chemistry that aqueous solutions always contain H^+, however little there might be. Write a chemical reaction showing CN^- behaving as a weak base in the presence of H^+.

15. From your knowledge of the pH of aqueous solutions, what happens to [H^+] as the pH of a solution becomes more acidic?

16. According to the reaction you wrote above showing CN^- acting as a base, as a solution containing CN^- becomes more acidic what would LeChâtelier's principle predict should happen to [CN^-]?

17. What would LeChâtelier's principle predict should happen to [HCN]?

18. Do these results agree with what's actually shown in Table 1?

19. In Q13 your group formulated an explanation for where the HCN in these solutions comes from. Was your explanation reasonable? If not, adjust it based on your improved understanding.

20. Suppose that a small amount of AgCN dissolves, yielding some Ag^+ and some CN^-. Suppose further that some of this CN^- is consumed by the reaction of CN^- with H^+ due to the basic character of CN^-. As a group, decide what we should expect will happen to [Ag^+] on the basis of LeChâtelier's principle.

21. Given your conclusion, what should we expect to see happen to [Ag^+] in saturated solutions of AgCN as the pH becomes more acidic?

22. With reference to Table 1, is this what actually happens?

23. Earlier you discovered that there are conditions under which the assumption $[Ag^+] = [CN^-]$ seems to work.

   a. Under what conditions of pH does this assumption appear not to work?

   b. Explain why this assumption doesn't work under these conditions.

24. In summary, what should happen to the solubility of AgCN as the pH of the solution is decreased (becomes more acidic)?

25. As a group, write one or two complete sentences that explain why the solubility of AgCN changes as a function of pH.

26. Explain how your understanding of the solubility of sparingly soluble salts has progressed over the approach you learned in General Chemistry.

Application

27. Lanthanum Fluoride, LaF_3, is an insoluble fluoride salt having $pK_{sp} = 18.7$ ($K_{sp} = 2 \times 10^{-19}$). Write chemical reactions showing the equilibrium reactions of LaF_3 dissolving and fluoride ion acting as a weak base in water. Would you expect the solubility of LaF_3 to be pH dependent the way AgCN is? Explain why or why not. If you think this solubility would be pH dependent, under what pH conditions would you expect LaF_3 to be most soluble?

28. Marble is a naturally occurring form of calcium carbonate (calcite, $CaCO_3$) which is widely used as a building material for exterior use. Calcite exhibits a K_{sp} of 4.5×10^{-9}. Carbonate ion, CO_3^{-2}, is a weak base in water (its conjugate acid is bicarbonate, HCO_3^-) having $pK_b = 3.671$. Would you expect marble to exhibit pH dependence in its water solubility? Why or why not? Buildings that are constructed of or covered with marble are regularly exposed to wetness from rain. What consequences does your expectation above imply for the durability of this material?

29. Heavy metal sulfides such as CdS and PbS are often significant components of mine tailings (the piles of waste and low-grade ores discarded by a mining operation). While these sulfides are quite insoluble in water (K_{sp} of CdS = 1×10^{-27}, K_{sp} of PbS = 3×10^{-28}), over the years rainwater and snowmelt form equilibrium solutions of these compounds and subsequently wash heavy metal ions down into streams and other natural waters where they constitute toxic pollutants. In recent years some concern has arisen over the lowering of the pH of rainwater ("acid rain") due to dissolution of acidic gases such as SO_2 and NO_2 in the rainwater. Would you expect this situation to reduce the pollution of natural water with heavy metals or make the problem worse? Justify your answer based on the solubility of salts of weak bases such as S^{-2}.

30. Thus far we have been picking exclusively on reactions involving CN^-, but there's no reason why we shouldn't give equal opportunity to Ag^+. For example, you're probably aware that some metal ions, like Ag^+, form insoluble hydroxides. Furthermore this process is going to be tough to avoid in strongly basic solutions. So consider the reaction

$$Ag_2O(s) + H_2O \rightleftharpoons 2Ag^+ + 2OH^- \qquad K_{sp} = 3.8 \times 10^{-16}$$

a. If Ag_2O forms to an appreciable extent what effect will this equilibrium have on the solubility of AgCN? Explain your reasoning. Judging by the equilibrium reaction shown, would you expect the effect of this equilibrium on the solubility of AgCN to be pH dependent?

b. Try this little experiment. Pick the solution from Table 1 where your reasoning suggests that the precipitation of Ag_2O will be most significant. You can easily figure out $[OH^-]$ from the pH and $[Ag^+]$ that is given you, so calculate the ion product suggested by the reaction above and determine whether or not precipitation of Ag_2O will actually occur.

31. Actually, things are even worse than this. Ag^+ can form an ion pair with OH^- to form $AgOH(aq)$ which can further react with another OH^- to form the complex ion $Ag(OH)_2^-$. Here are the equilibria:

$$Ag^+ + OH^- \rightleftharpoons AgOH(aq) \quad K_{f1} = 100$$

$$AgOH(aq) + OH^- \rightleftharpoons Ag(OH)_2^- \quad K_{f1} = 100$$

If these equilibria happened to an appreciable extent, what effect will it have on the solubility of $AgCN(s)$? Justify your answer with appropriate reasoning. Will you expect the effect of these equilibria to be pH-dependent?

pH of Solutions of Strong Acids and Bases

Learning Objectives

Students should be able to:

Content

- Calculate the pH of solutions containing any concentration of strong acid or base.

Process

- Recognize when the result of a calculation makes sense (Critical thinking)

- Identify assumptions. (Problem solving)

Prior knowledge

- Be familiar with the dissociation of strong acids and bases in water at the General Chemistry level

- Be familiar with the concept of pH and the autoionization of water at the General Chemistry level.

- Be familiar with the concept of chemical activity and its ionic strength dependence, and be able to use activity coefficients to determine the activity of an ion having known molar concentration.

Further Reading

- Harris, D.C. 2010. *Quantitative Chemical Analysis,* 8th Edition, Section 8-1, p.163-165. New York: W.H. Freeman.

Authors

David L. Langhus and Juliette M. Lantz

Consider this...

The pH of any solution can be calculated from

$$pH = -\log_{10} [H^+]$$

Hydrochloric acid, HCl, is a strong acid in water.

Key Questions

1. 1.00×10^{-3} moles of HCl is dissolved in a liter of water. Calculate $[H^+]$ in this solution. What is the pH of this solution?

2. Think with your group about how you obtained these results and agree on how one could obtain the pH of any HCl solution in water given its initial concentration.

3. Split up the following among your group and calculate the pH of each HCl solution. When you have a value for each, look them over as a group to be sure you agree. If there are concentrations or ranges of concentration for which your group isn't able to agree be prepared to explain what the problem is. You've already done 1.00×10^{-3} M so just fill it in:

Initial Conc. HCl, M	[H+]	pH
0.100		
0.0100		
1.00×10^{-3}		
1.00×10^{-4}		
1.00×10^{-5}		
1.00×10^{-6}		
1.00×10^{-7}		
1.00×10^{-8}		
1.00×10^{-9}		
1.00×10^{-10}		
1.00×10^{-11}		
1.00×10^{-12}		

Table 1 *The hydrogen ion concentrations and pH of several aqueous hydrochloric acid solutions.*

4. As a group consider all the pH values calculated in Table 1. Do any of the values your group obtained seem inconsistent with your intuitive understanding of the pH of solutions in water? If so agree among yourselves about the concentration range(s) over which this inconsistency arises and explain why you think it's inconsistent.

5. The autoionization constant for water, K_w, is 1.00×10^{-14} at 25°C. What is $[H^+]$ in pure water at 25°C without any HCl or anything else in it at all, and where does this H^+ come from since there's no acid present?

6. What is the smallest concentration of H^+ you would expect to observe in any solution of HCl in water, no matter how dilute? Agree among your group on this value and explain why you think it must be so.

7. On the basis of your answer above, what is the most basic pH value you'll expect to observe in any solution of HCl in water?

Consider this...

Table 2 shows the actual pH values measured using a glass electrode for some carefully prepared solutions of HCl in pure water.

Table 2 pH of some solutions of HCl in pure water	
Initial HCl Conc. M	**pH**
0.10	1.08
0.010	2.04
0.0010	3.01
1.0×10^{-4}	4.00
1.0×10^{-5}	5.00
1.0×10^{-6}	6.00
1.0×10^{-7}	6.79
1.0×10^{-8}	6.98
1.0×10^{-9}	7.00
1.0×10^{-10}	7.00
1.0×10^{-11}	7.00
1.0×10^{-12}	7.00

Key Questions

8. Examine the pH values in Table 2. What happens to the pH of these solutions as less and less acid is added?

9. Compare the pH values you calculated for Table 1 in Q3 with the actual values shown in Table 2. Are there concentration ranges where there appears to be a problem with the way you calculated some of the values? If so, agree on the concentration ranges over which this happens and indicate what it is about the values you calculated in Q3 that tips you off to this problem.

10. pH is ordinarily measured using a pH electrode which responds to the *activity* of H^+ rather than its actual concentration. Recall that activity is related to concentration by the relationship

$$A_H = \gamma[H^+]$$

where A_H is the activity of H^+ and γ is the activity coefficient for H^+. The pH is then more properly calculated using

$$pH = -\log_{10} A_H$$

Here's a table of activity coefficients for H^+.

Table 3 Activity coefficient, γ, for H^+ in water at 25°C (obtained from EDH)	
Ionic Strength, M	γ
0.10	0.825
0.010	0.914
0.0010	0.967
1.0×10^{-4}	0.989
1.0×10^{-5}	0.996

You've already calculated $[H^+]$ for 0.100 M HCl in Table 1, so use that to calculate the activity of H^+, and from that, the pH.

11. Reach group consensus regarding the pH range over which taking activity into account would improve the agreement between your calculations in Q3 and the actual pH values in Table 2. Be prepared to explain your reasoning.

12. In any solution of HCl in water we'd expect to find not only H^+ but also Cl^-. What concentration of Cl^- would you expect to observe in a solution of HCl with an initial concentration of 1.00×10^{-8} M in water?

13. In a solution containing water as the solvent we'd also expect to find some concentration of OH^-. Write a mathematical expression that would permit you to calculate $[OH^-]$ if you knew $[H^+]$.

14. If you were to handle a bottle containing an HCl solution in water that was initially 1.00×10^{-8} M, it is unlikely that you would get an electrical shock from it, so it must be that the number of positive ions in the bottle exactly equals the number of negative ions. Write a mathematical expression representing this equality that takes into account all the ions present in a 1.00×10^{-8} M solution of HCl in water. You obtained a value for $[Cl^-]$ and an expression for $[OH^-]$ above, so plug those in too.

If you clear the fractions in this equation by multiplying through by any denominator you'll discover that this gives you a quadratic equation in $[H^+]$. You can rearrange some things to get it into the form

$$a[H^+]^2 + b[H^+] + c = 0$$

where a, b and c are coefficients for which you probably have actual numerical values. (The value of a is probably 1.00 in this case so don't let that hang you up.) You can get a value for $[H^+]$ from such an equation by using the quadratic formula

$$[H^+] = \frac{-b \pm \sqrt{b^2 - 4ac}}{2a}$$

Actually, you are likely to arrive at two different values for $[H^+]$ using this equation. One of them is typically negative or otherwise bears a value that is physically impossible.

15. Solve your equation for [H$^+$]. What result does this approach give for the pH of a 1.00×10^{-8} M HCl solution in water? Does this value seem more reasonable than that which you obtained in Q3? If so, in what way is it more reasonable?

16. Use the approach in Q12-Q14 to calculate the pH of a solution of HCl in water that's initially 1.00×10^{-3} M. Is the result reasonable based on your chemical intuition?

17. Try this approach to obtain the pH of a solution of HCl in water that's initially 1.00×10^{-10} M. Is the result reasonable based on your chemical intuition? Be prepared to explain why you think it's reasonable or not.

Try Q18-Q20 on your own. Then get together with your group and reach consensus on them:

18. For a solution of HCl in water that's initially 1.00×10^{-3} M what appears to be the primary source of H^+?

    Mainly the acid   Both acid and water contribute somewhat   Mainly the water

19. For a solution of HCl in water that's initially 1.00×10^{-8} M what appears to the primary source of H^+?

    Mainly the acid   Both acid and water contribute somewhat   Mainly the water

20. For a solution of HCl in water that's initially 1.00×10^{-10} M what appears to be the primary source of H^+?

    Mainly the acid   Both acid and water contribute somewhat   Mainly the water

21. On the basis of what you've learned thus far regarding accurately calculating the pH of HCl solutions, you ought to be able to divide the HCl concentration range into several regions corresponding to the method you would use to calculate the pH of HCl solutions as easily as possible. Reach group consensus on how many such regions there are (typically groups arrive at three or four) and for each region indicate what method you'd use to do the calculation and what assumptions, if any, you'd be making as you did so.

Region 1:

Region 2:

Region 3:

Region 4:

22. Now think back to Q3 where you originally attempted to calculate the pH of all those HCl solutions. List the assumptions you were making when you did that. Is there any pH range over which these assumptions appear to hold up?

23. Why do you think it's important to identify the assumptions being made when one calculates equilibrium concentrations?

Applications

24. Calculate the pH of an HCl solution in pure water that's initially 5.0×10^{-8} M. Does the value you obtain appear to follow the trend you see in Table 2?

25. Calculate the pH of an aqueous solution that's initially 5.0×10^{-8} M in HCl and 0.10 M in NaCl. Don't forget that activities need to be used in the K_w expression but actual concentrations are called for when balancing positive and negative charges.

26. Calculate the pH of a solution of NaOH in pure water that's initially 0.10 M. Identify any assumptions you make.

27. Calculate the pH of a solution of NaOH in pure water that's initially 1.0×10^{-7} M.

28. Calculate the pH of a solution of NaOH in pure water that's 1.0×10^{-10} M, indicating any assumptions you make.

29. A wealthy benefactor bankrolls the construction of a brand-new swimming pool for a struggling kids' camp. After a few months of mayhem, the bulldozers, masons, plumbers, etc. leave behind a sparkling facility filled with 100,000 gallons of well water which, in this locale, is primarily rainwater so it contains next to nothing in the way of dissolved salts of any kind. Everything is peachy keen for about a week, but then the newly trained lifeguard comes in with the bad news that the pH of the water has steadily risen to 8.0, well outside the county's requirement of 7.2–7.8. (Rises in pH with time are normal in pools chlorinated with hypochlorite solution due to the NaOH in which the hypochlorite is delivered, but brand-new pool owners are rarely aware of this.) A call to the local home pool and spa store results in the laconic instruction "add some muriatic acid (12 M HCl)," but this isn't entirely helpful because the guy on the phone claims he can't tell them how much to add. Muriatic acid sounds like a chemical and it really stinks so one certainly wouldn't want to add very much of it to water kids are going to swim in.

    Proposal A: Take the muriatic acid bottle to the kitchen and dilute a teaspoon of it to a gallon (making a solution that's initially 0.015 M HCl). This makes it not stink anymore and, in fact, it doesn't even burn but it still tastes awful. (Don't try this at home.) Suppose we put 1/8 cup of this in the pool (making it about 1.0×10^{-9} M in HCl). A teen staffer who's just finished AP chemistry in high school objects because 10^{-9} M HCl in the pool will make the pH 9 and then it'll be even worse than it is now. Is he right? What assumptions is he making?

    Proposal B: The teen staffer suggests adding more of the 0.015 M HCl. Assuming that the pool contains 4.0×10^5 L of pure water, it should take 1.7 L of this dilute acid to get the initial HCl concentration to 6.3×10^{-8} M and the pH to 7.2. If the pool did indeed contain pure water, what pH would he actually end up at using this method? Would it be an improvement from a regulatory point of view?

    Proposal C: Nobody can come to agreement on this, so they call you on the phone and explain the dilemma. Of course you know that the pH is 8.0 because some NaOH has been added. Estimate how many moles of NaOH are in there.

Calculate carefully the pH of the pool upon the addition of 25 mL of the original muriatic acid (12 M HCl) if the pool otherwise contains just this amount of NaOH in pure water.

Another detail to consider:

If you care for a swimming pool or spa you know that this whole thing is bogus because pool water is saturated with CO_2, a decent weak acid that forms an equally decent buffer upon addition of a little base, not to mention the pounds of carbonate as $NaHCO_3$ or Na_2CO_3 that commercial swimming pool operators add. A pool this size is going to need about a gallon of muriatic acid to budge the pH appreciably in the direction these folks need to go.

Acid-Base Distribution Plots

Learning Objectives

Students should be able to:

Content

- Describe how the molar concentrations of mono- and polyprotic weak acids and their conjugate bases vary with pH.

- Identify the principal species resulting from the dissociation of a weak acid at a given pH.

Process

- Sketch ionic distribution graphs given acid-base parameters. (Critical thinking – visualization)

- Interpreting graphs. (Information processing)

Prior knowledge

- Understand how weak acids dissociate and the concept of K_a for mono- and polyprotic acids at the General Chemistry level, including the relationship of K_a to acid strength.

- Be able to translate between K_a and pK_a.

Further Reading

- Harris, D.C. 2010. *Quantitative Chemical Analysis*, 8th Edition, Sections 8-3 and 8-4, p.166-171. New York: W.H. Freeman.

- Skoog, D.A., D.M. West, F.J. Holler, and S.R. Crouch. 2004. *Fundamentals of Analytical Chemistry*, 8th Edition, Section 15H, p. 419. Belmont, CA: Thomson Brooks/Cole.

Authors

David L. Langhus and Juliette Lantz

PGLANA004018a POGIL

Consider this...

Suppose that we prepare several 0.0100 M solutions of acetic acid, each having a different pH*. The following equilibrium is established in each case:

$$HAc \rightleftharpoons H^+ + Ac^- \qquad pK_a = 4.76$$

Table 1 shows the results we'd get if we measured the molar concentration of acetate ion in each of these solutions using, for example, an acetate sensitive electrode**:

Table 1 Acetate concentrations** in acetic acid solutions having various pH values	
pH	**[Ac⁻], M**
13.00	0.0100
11.00	0.0100
9.00	0.0100
8.00	0.0100
7.00	9.96×10^{-3}
6.00	9.58×10^{-3}
5.00	6.93×10^{-3}
4.00	1.84×10^{-3}
3.00	2.21×10^{-4}
2.00	2.25×10^{-5}
1.00	2.26×10^{-6}

* One way to set the pH of a solution at a constant value is to use a pH buffer. For the case of this activity buffers have been used that don't contain acetic acid or acetate.

** An ion sensitive electrode responds to chemical activity rather than concentration. For Table 1, the activities obtained have been divided by the appropriate activity coefficient to give molar concentrations. Be aware of this if you attempt to compare these results with some you might calculate ignoring activity.

Key Questions

1. On the graph below, plot the acetate ion concentration as a function of pH. Compare your graph with those of the others in your group and come to agreement on how this should look.

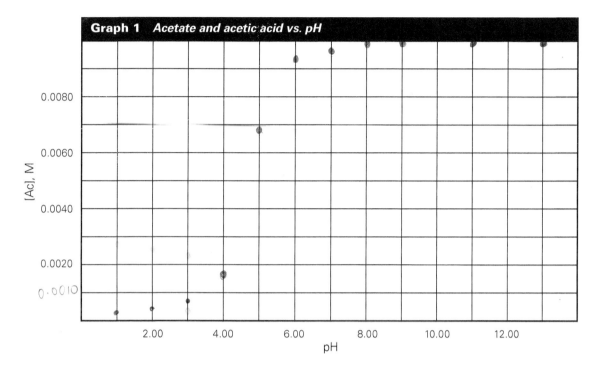

Graph 1 *Acetate and acetic acid vs. pH*

2. Recall that we put only acetic acid in each solution for Table 1. In the pH 5.00 solution, for instance, decide with your group where this acetate comes from and write your group consensus.

3. Using the initial concentration of HAc, 0.0100 M, and the appropriate [Ac⁻] from Table 1, calculate the molar concentration of acetic acid, [HAc], in that pH 5.00 solution.

$$[AC] = 6.93 \times 10^{-3}$$

$$[AC] + [HAC] = 0.0100$$

$$[HAC] = \boxed{3.07 \times 10^{-3}}$$

4. Divide the solutions below among the members of your group. For each solution calculate [HAc] and enter the values in Table 2 below. You've already done the pH=5.00 case, so just fill that in. Once your group agrees on the values, plot them on Graph 1. Label the two lines with [Ac⁻] and [HAc] as appropriate.

Table 2 Acetate and acetic acid concentration in solutions of acetic acid having various pH values		
pH	**[Ac⁻], M**	**[HAc], M**
13.00	0.0100	◯
11.00	0.0100	◯
9.00	0.0100	◯
8.00	0.0100	◯
7.00	9.96×10^{-3}	$.4.00 \times 10^{-5}$
6.00	9.58×10^{-3}	4.2×10^{-4}
5.00	6.93×10^{-3}	3.07×10^{-3}
4.00	1.84×10^{-3}	8.16×10^{-3}
3.00	2.21×10^{-4}	9.78×10^{-3}
2.00	2.25×10^{-5}	9.98×10^{-3}
1.00	2.26×10^{-6}	1.00×10^{-3}

5. Examine Graph 1. At what pH do the lines for [HAc] and [Ac⁻] cross?

$pka = pH$

$\boxed{4.76 = 4.76}$

Consider this…

A solution is prepared which is initially 1.00 M in NaHSO$_4$:

$$HSO_4^- \rightleftharpoons H^+ + SO_4^{2-} \quad pK_a = 1.99$$

Here's what happens to the equilibrium concentrations of sulfate-containing ions as we adjust the pH.

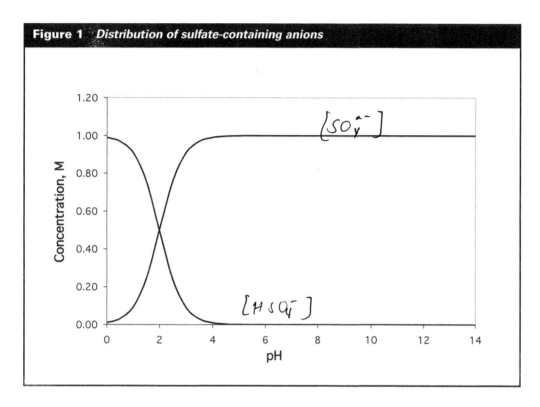

Figure 1 *Distribution of sulfate-containing anions*

[SO$_4^{2-}$]

[HSO$_4^-$]

Key Questions

6. Label the lines in Figure 1 with [HSO$_4^-$] and [SO$_4^{2-}$] as appropriate. At what pH do the lines cross in this case?

pH = 1.99

7. Come to consensus in your group regarding the mathematical relationship between [HSO$_4^-$] and [SO$_4^{-2}$] at the pH where the lines cross. Write down this relationship and justify it.

It has an inverse relationship, because as the other goes up the other one goes down.

8. Write the K_a expression corresponding to the HSO_4^- – SO_4^{2-} equilibrium. Use the mathematical relationship you obtained above to simplify this expression at this particular value of $[H^+]$.

$$k_A = \frac{[H^+][SO_4^{2-}]}{[HSO_4^-]} \longrightarrow k_A = [H^+]$$

9. As a group, brainstorm a way to predict the pH at which these lines cross.

use the Henderson - Hasselbach equation, when the pka is given.

$$pH = pka + log \frac{base}{acid}$$

10. Consider the acetic acid case you examined earlier in Graph 1. Can the analysis your group did of the line crossing pH in Q7 and Q8 be extended to include the acetic acid case? Does the algorithm your group proposed in Q9 also work for acetic acid? If not, propose a modification so that your algorithm works more generally.

Yes, it can.

11. Which of HSO_4^- and SO_4^{2-} is most abundant (the principal species) at pH 4.00? What is the principal species at pH 1.50?

HSO_4^-

SO_4^{2-}

Consider this...

Here's the distribution of a few weak acids and their conjugate bases:

Figure 2

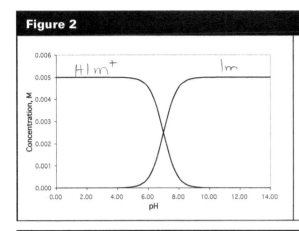

0.0050 M imidazole (Im)
and its conjugate acid, HIm⁺

$$HIm^+ \rightleftharpoons H^+ + IM$$
$$pK_a = 6.99$$

Figure 3

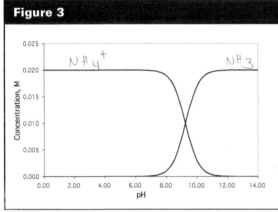

0.020 ammonia
and ammonium ion

$$NH_4^+ \rightleftharpoons H^+ + NH_3$$
$$pK_a = 9.24$$

Graph 2

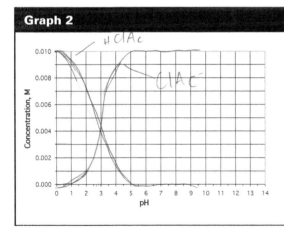

0.010 M chloroacetic acid (HClAc)
and its conjugate base ClAc⁻

$$HClAc \rightleftharpoons H^+ + ClAc^-$$
$$pK_a = 2.865$$

Key Questions

Divide your group roughly in half, assigning Q12 to one half and Q13 to the other half. Once you have answers, examine them as a group to be sure you agree:

12. Examine Figure 2, the distribution of imidazole and its conjugate acid. Which of HIm^+ and Im is the principal species present at pH 4.00? At pH 8.00?

HIm$^+$ is the principal @ pH 4.00.
Im is the principal @ pH 8.00.

13. Examine Figure 3, the distribution of ammonia and ammonium ion. Which of NH_4^+ and NH_3 is the principal species present at pH 4.00? At pH 8.00?

NH_4^+ principal in pH 4.00
NH_3 principal in pH 8.00.

14. At what pH do the lines cross in each of these graphs? Is this consistent with the results your algorithm from Q10 yields?

Fig 1 → pH 6.99
Fig 2 → pH . 9.24

15. On the grid provided for Graph 2, sketch the distribution graph you'd expect for chloroacetic acid and its conjugate base as a function of pH in 0.010 M chloroacetic acid. **Don't do any calculations.** Just sketch it. Compare your result with the others in your group and as a group, settle on one representation that you think most accurately describes the behavior of this acid. Label the two lines with their respective species so you can tell them apart.

16. What is the principal species present in solution at pH 1.00? At 4.00?

HClAc   ClAc$^-$

17. Use your sketch to estimate the molar concentration of the chloroacetate ion at pH 8.00.

18. Now compare all three of Figures 1 and 2 and Graph 2. Critique this statement individually: "For any weak acid, the acid form is the dominant species at pH values more acidic than 7.00 and the conjugate base is dominant at pH values basic of pH 7.00." Support or refute this statement based on the distributions diagrams you've seen.

We refute this statement because, based on the graphs, the pH at which they cross differs for each. It doesn't always cross at pH 7.00.

19. Share your answer to the previous question with your group and agree with your group on the best answer.

20. Notice that for the chloroacetic acid–chloroacetate conjugate acid-base pair the pH range over which chloroacetic acid is the principal species is quite a bit narrower than that for the chloroacetate ion. Now consider a collection of weak acids with pK_a values between about 1 and 10. How does the breadth of the pH range over which the acid form is dominant vary in comparison with that for the conjugate base form?

Consider this...

Some weak acids exhibit more than one acidic proton. Consider, for example, carbonic acid, H_2CO_3, the acid formed by dissolving CO_2 (g) in water:

$$H_2CO_3 \rightleftharpoons H^+ + HCO_3^- \qquad pK_{a1} = 6.35$$

Once that first proton is gone to form bicarbonate there's still another that can be lost, but as you might imagine it's a lot tougher to pull that second one off. (We'd say that the carbonate anion, CO_3^{2-}, is a fairly strong base.)

$$HCO_3^- \rightleftharpoons H^+ + HCO_3^{2-} \qquad pK_{a1} = 10.33$$

This means that we can have as many as three species floating around in solution all at once: H_2CO_3, HCO_3^- and CO_3^{2-}. Here's how they distribute in a 0.0050 M carbonic acid solution:

Figure 4 *Distribution of carbonate-containing species*

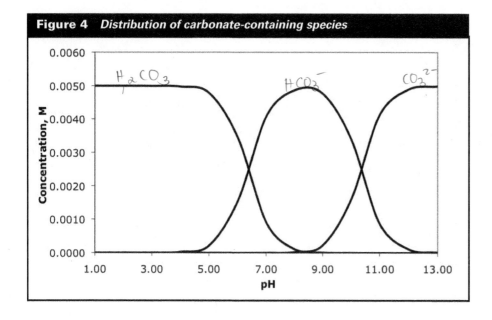

Key Questions

21. Label the lines in Figure 4 with [H_2CO_3], [HCO_3^-] and [CO_3^{2-}], as appropriate. At what pH do the H_2CO_3 and HCO_3^- lines cross? How about the HCO_3^- and CO_3^{2-} lines? How could you have predicted this?

pH = 6.35

pH = 10.33

based on the given pkas.

(pH = pKA

22. Identify the most abundant and the second most abundant species at each of pH 6.00, 7.00, and 10.50. Write your results in the table below:

Table 3	Species hierarchy for carbonic acid at several pH values	
pH	**Most Abundant**	**Second Most Abundant**
6.00	H_2CO_3	HCO_3^-
7.00	HCO_3^-	CO_3^{2-}
10.50	CO_3^{2-}	HCO_3^-

Consider this...

Consider the dicarboxylic acid phthalic acid. For simplicity, we'll represent it as H_2Ph.

$$H_2Ph \rightleftharpoons H^+ + HPh^- \quad pK_{a1} = 2.950$$

$$HPh^- \rightleftharpoons H^+ + Ph^{2-} \quad pK_{a2} = 5.408$$

In this case, we might observe any or all of H_2PH, HPh^- and Ph^{2-}. Here's how they distribute in a 0.0010 M phthalic acid solution.

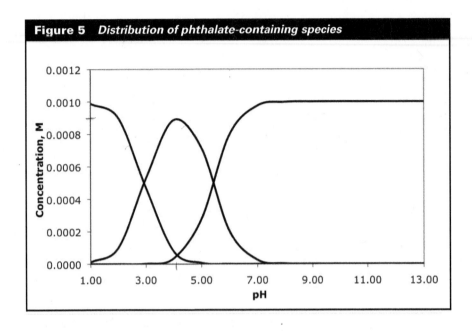

Figure 5   Distribution of phthalate-containing species

Key Questions

23. Over what pH range is HPh^- the dominant species in phthalic acid solutions?

pH range between   2.950 - 5.408

24. Estimate the molar concentration of HPh^- at pH 4.00.

~ 0.0093   pH

25. In the distribution diagrams for carbonic and phthalic acid (Figures 4 and 5), does each species reach the initial concentration at some pH?

26. With your group, propose acid characteristics that might lead to a species not ever reaching the initial concentration at any pH.

27. Brainstorm with your group the benefits of using distribution graphs to represent acid-base systems. Record your insights below. Extra credit awarded for using the word "visualize" in your answer.

28. On a scale of 1-5, rate your comfort level with sketching a useful graph given a few pKa values. Explain why you chose this particular level.

Applications

29. Write a chemical equation showing the equilibrium between hydroxylamine, $HONH_2$, and the hydroxylammonium ion, $HONH_3^+$.

Draw the distribution graph for a 0.010 M solution of hydroxylamine.

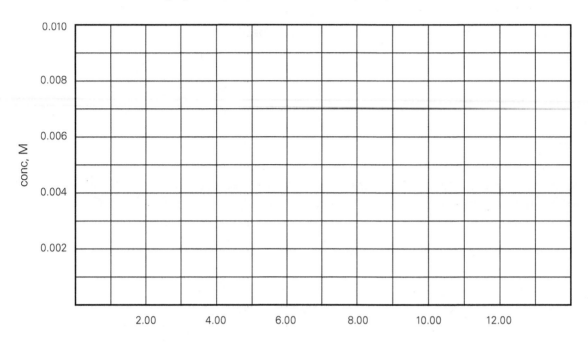

What is the principal nitrogen-containing species at pH 7?

Over what pH range is this compound primarily present as $HONH_2$?

30. Although we've thus far characterized the dissociation of, for example, acetic acid in terms of the molar concentrations of HAc and Ac⁻, it's also common to describe this using the percent of the total acetic acid present as HAc and the percent as Ac⁻. Clearly, since the acetate ion has to be in one form or the other, the sum of the percent HAc and the percent Ac⁻ is always 100%.

Determine the percentage of HAc and percentage of Ac⁻ in the solutions prepared for Table 1 and graph percentage of HAc and percentage of Ac⁻ as functions of pH.

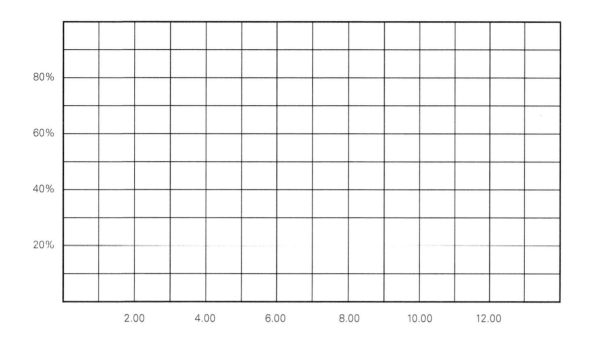

Compare the graph above with that which you made for Q1 and Q5. Do the lines cross at the same pH? Should they?

31. Orthophosphoric acid is a tri-protic acid that undergoes the following equilibria in water:

$$H_3PO_4 \rightleftharpoons H^+ + H_2PO_4^- \qquad pK_{a1} = 2.148$$

$$H_2PO_4^- \rightleftharpoons H^+ + HPO_4^{2-} \qquad pK_{a2} = 7.199$$

$$HPO_4^{2-} \rightleftharpoons H^+ + PO_4^{3-} \qquad pK_{a3} = 12.15$$

You may plot the concentrations of the various phosphate-containing ions below if you'd like.

a. What is the principal phosphate-containing species at pH 10.00?

b. Over what pH range is the $H_2PO_4^-$ ion dominant?

c. Within what range would one have to adjust the pH of a solution to be sure that most of the phosphate present was actually in the PO_4^{3-} form in water?

32. The amino acid alanine exhibits two functional groups: a carboxylic acid and an amine. When the amine group is protonated alanine is a diprotic acid. Here are the equilibria in water:

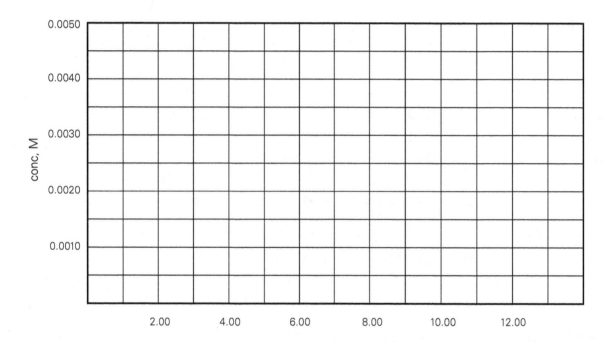

$$H_3C–CH–\overset{\overset{\displaystyle O}{\|}}{C}–OH \rightleftharpoons H^+ + H_3C–CH–\overset{\overset{\displaystyle O}{\|}}{C}–O^- \qquad pK_{a1} = 2.348$$

(below first structure's CH is) NH_3^+ ... NH_3^+

(zwitterion)

$$H_3C–CH–\overset{\overset{\displaystyle O}{\|}}{C}–O^- \rightleftharpoons H^+ + H_3C–CH–\overset{\overset{\displaystyle O}{\|}}{C}–O^- \qquad pK_{a2} = 9.867$$

NH_3^+ ... NH_2

Alanine is the neutral species (shown as the zwitterion above) which we might represent as HA. The fully protonated form would then be written H_2A^+, making it look just like any other diprotic acid:

$$H_2A^+ \rightleftharpoons H^+ + HA \qquad pK_{a1} = 2.348$$
$$HA \rightleftharpoons H^+ + A^- \qquad pK_{a2} = 9.867$$

Suppose that we prepare a solution of alanine having initial concentration 0.0050 M. Draw the distribution graph below.

[distribution graph: y-axis "conc, M" from 0.0010 to 0.0050; x-axis from 2.00 to 12.00]

a. Draw the principal form of alanine at pH 2.00. At pH 10.00.

b. Over what pH range is the zwitterion dominant? Is there any pH at which its concentration actually reaches 0.0050 M?

c. The *isoelectric pH* of an amino acid is that pH at which the zwitterion exhibits maximum concentration. Estimate the isoelectric pH of alanine.

33. A potential concern when preparing solutions like those in Table 1 is that the buffers used have sufficient capacity to resist significant pH change upon the addition of the HAc added in each case. So let's pick on the pH 7.00 buffer. Assume that the pK_a of the buffer's acid component is equal to its nominal pH of 7.00.

Suppose that we use 0.10 M buffers (that is, the sum of the concentrations of the buffer acid and its conjugate base is 0.10 M) and that we prepare the solutions in Table 1 by adding 1.00 mL of 1.00 M acetic acid to 99.00 mL of buffer.

a. How many moles of the buffer acid, HB, and its conjugate base, B⁻, are present?

b. In the worst case, if the H⁺ from all of the acetic acid added were consumed by the buffer, how many moles of each buffer component would remain?

c. Would this change the pH much? You can make use of the Henderson-Hasselbalch equation if you're not sure.

The Acid-Base Distribution Functions

Learning Objectives

Students should be able to:

Content

- Explain what information acid-base fractional distribution functions provide.

- Use acid-base distribution functions to calculate the concentration of a species arising from any weak acid at a given pH.

Process

- Extend the application of a method to new situations using pattern recognition. (Critical thinking)

- Work effectively with others to reach a consensus. (Teamwork)

- Evaluate answers for reasonableness. (Critical thinking)

Prior knowledge

- Be able to write and use K_a expression for mono- and polyprotic acids.

- Be able to write mass balance expressions.

Further Reading

- Harris, D.C. 2010. Quantitative Chemical Analysis, 8th Edition, Section 9-5, p. 197-198. New York: W.H. Freeman.

Author

David L. Langhus

Consider this...

Several acetate solutions are prepared in 0.1 M buffers. Each solution contains 0.0020 M acetic acid, HAc, and 0.0010 M sodium acetate, NaAc. The solutions are thoroughly mixed and allowed to come to equilibrium. Some of the equilibrium concentrations of acetic acid ([HAc]) and acetate ion ([Ac–]) are then measured with the following results:

Table 1 *Molar concentrations of acetate ion and acetic acid as a function of pH*

pH	[Ac-], M	[HAc], M
3.00	0.0001	0.0029
4.00	0.0005	0.0025
5.00		
6.00		0.0002
7.00	0.0030	0.0000

Key Questions

1. You'll notice that acetate comes from two places in the solutions above. Some is added as sodium acetate, which dissociates completely under these conditions as we'd expect. The rest comes from acetic acid that may not dissociate completely, depending upon the pH. What is the total concentration of acetate, without regard for what it may be bound to, in each of these solutions?

2. Examine the column in Table 1 containing concentrations of HAc. What happens to [HAc] as the pH becomes more basic? Agree with your group members on a justification for this behavior.

3. Only 0.0020 M HAc was initially put into each solution in Table 1. So why does [HAc] have the value it has at pH 3.00? Where does the additional HAc come from?

4. Only 0.001 M NaAc was put into each solution initially. Does the value of [Ac⁻] at pH 7.00 make sense compared to the other values in Table 1? Be prepared to explain your reasoning.

5. Divide up the blanks in Table 1 and use the K_a expression for the acetic acid equilibrium or the Henderson-Hasselbalch equation, as you prefer, to calculate the appropriate quantity. The pK_a for acetic acid is 4.76. Do the concentrations you get seem to make sense?

A general monoprotic weak acid, HB, might dissociate thus:

$$HB \rightleftharpoons H^+ + B^-$$

Although B might be originally added to the solution as HB or B⁻ (as in NaB), suppose that neither HB nor B⁻ participate in any reactions other than this equilibrium, so the total concentration of B from whatever source is simply [HB] + [B⁻].

The symbol is defined as the fraction of all the HB and B⁻ present ([HB] + [B⁻]) that's actually in the B⁻ form. So if half of all the HB and B⁻ in the solution is in the B⁻ form, then α_{B^-} is 0.5. Here's a more formal mathematical definition:

$$\alpha_{B^-} = \frac{[B^-]}{[HB] + [B^-]}$$

Table 2 *Molar concentrations of acetate ion and acetic acid, along with corresponding α values, as a function of pH*

pH	[Ac⁻], M	[HAc], M	α_{Ac^-}	α_{HAc}
3.00	0.0001	0.0029		
4.00	0.0005	0.0025	0.17	0.83
5.00				
6.00		0.0002		
7.00	0.0030	0.0		

Key Questions

6. The first three columns of Table 2 are just Table 1, so fill in the blanks in columns 2 and 3 with the results you've already calculated for Table 1. Then split the solutions in Table 2 among the members of your group and for each calculate α_{Ac^-}, the proportion of acetate present that's actually in the Ac⁻ form. You've already calculated the value of [HAc] + [Ac⁻] in all of these solutions. Look over the results you end up with as a group to be sure the values trend the way you think they should.

7. A solution of acetic acid, HAc, is prepared having an initial concentration of 5.0×10^{-2} M. The resulting solution is treated with a buffer such that $\alpha_{Ac^-} = 0.65$. What is the molar concentration of acetate in this solution, assuming that its only source is the acetic acid? Show the mathematical expression for you used to obtain this result.

α_{HB}, the fraction of all the HB and B$^-$ in solution that's HB, is defined as:

$$\alpha_{HB} = \frac{[HB]}{[HB] + [B^-]}$$

Key Questions

8. As before, distribute among yourselves the solutions in Table 2 and calculate α_{HAc} for each. Do the results trend in the way you expect?

9. For each pH value add α_{Ac^-} and α_{HAc}. What do you notice about the sum in each case? With your group, come up with a justification for this behavior.

Consider a solution of the hypothetical monoprotic weak acid HB and its sodium salt, NaB. The relevant acid-base equilibrium would be written:

$$HB \rightleftharpoons H^+ + B^-$$

---

10. Write the mathematical expression for the K_a of this system and a mass balance expression for the sources and fates of B^-.

11. Let's get rid of the variable [HB] in the K_a expression by solving the mass balance expression for [HB] and making this substitution into the K_a expression. Then solve the K_a expression for [B⁻].

12. Recalling that

$$\alpha_{B^-} = \frac{[B]}{[HB] + [B^-]}$$

obtain a formula for α_{B^-} in terms of K_a and [H⁺]. Compare the result you get with those of the rest of the members of your group. You'll likely discover more than one formula. Make sure that they all amount to the same thing mathematically.

13. Use your formula to calculate α_{Ac^-} for acetic acid (pK_a= 4.76) at pH 5.00. Check with the other members of your group to be sure that you're all getting the same value, even though the formula each of you used might be different. Also compare what you get with the value you computed in Table 2. Do they agree?

14. Use the expression of your choice above to obtain a formula for α_{HB} in terms of K_a and $[H^+]$. You might review the result you got in Q9 in case it might help. Compare your result with those of the other members of your group and be sure that they all amount to the same thing mathematically.

15. Calculate α_{HAc} for acetic acid ($pK_a = 4.76$) at pH 5.00. Agree with the rest of your group on the numerical value and compare it with the result you calculated earlier for Table 2.

Consider this…

Carbonic acid (H_2CO_3) and phthalic acid are neutral examples of diprotic acids, and the fully protonated form of alanine can be regarded as H_2B^+, which behaves the same way. Consider a neutral diprotic acid H_2B. In a solution of H_2B, there's the possibility of three different B-containing species: H_2B, HB^- and B^{2-}. There are also three ways we can add "B" to the solution: as the free acid H_2B, the half salt NaHB (sodium bicarbonate and potassium hydrogen phthalate are common examples) and the full salt Na_2B. When preparing a B-containing solution, we may employ any one, two or all three of these compounds, depending upon what we're trying to accomplish.

It's no sweat to write the two K_a expressions and a mass balance expression for such a general solution. When you solve them the way you did the monoprotic HB case earlier you end up with the following distribution functions:

$$\alpha_{H_2B} = \frac{[H^+]^2}{[H^+] + K_{a1}[H^+] + K_{a1}K_{a2}}$$

$$\alpha_{HB^-} = \frac{K_{a1}[H^+]}{[H^+] + K_{a1}[H^+] + K_{a1}K_{a2}}$$

$$\alpha_{B^{2-}} = \frac{K_{a1}K_{a2}}{[H^+]^2 + K_{a1}[H^+] + K_{a1}K_{a2}}$$

These look pretty formidable, part of the reason you were not asked to derive them right here. The reason they're worthwhile is because they work for any diprotic acid. Not just phthalic acid, not just alanine, not just ethylenediammonium ion, but any diacid at all. They're all you have to know. You might consider it quite an undertaking to calculate α_{HB^-} at, say, pH 7.0, and it will take some button-pushing on your calculator, but consider the alternative: solving three simultaneous equations in three unknowns every time you need to know $[HB^-]$ at pH 7.0. That's what you've had to do until now. The Henderson-Hasselbalch equation wouldn't help you much in this case.

Key Questions

16. Calculate the molar concentration of KHP (potassium hydrogen phthalate, the potassium half-salt of the phthalate dianion) in a solution prepared by dissolving 100 millimoles of KHP in a liter of pH 5.00 buffer. For phthalic acid $pK_{a1} = 2.95$ and $pK_{a2} = 5.41$. Consider calculating the value of the denominator separately and making a note of its value in case it comes in handy in the next question. Each member of your group ought to get the same result. If you don't, figure out where the problem is so that you all agree.

17. Calculate the molar concentration of the fully protonated phthalic acid in this solution. It ought to be a lot easier to agree on a value for this.

Consider this...

So what about something like orthophosphoric acid, H_3PO_4? Looks like it's an example of a tri-protic acid, like H_3B. Now there are four different B^- containing forms with which to deal: H_3B, H_2B^-, HB^{2-} and B^{3-}. Here are a couple of the distribution functions.

$$\alpha_{H_3B} = \frac{[H^+]^3}{[H^+]^3 + K_{a1}[H^+] + K_{a1}K_{a2}[H^+] + K_{a1}K_{a2}K_{a3}}$$

$$\alpha_{H_2B^-} = \frac{K_{a1}[H^+]^2}{[H^+]^3 + K_{a1}[H^+] + K_{a1}K_{a2}[H^+] + K_{a1}K_{a2}K_{a3}}$$

This is where we ran out of gas. You don't actually think these things are fun to type do you? But that doesn't need to be a problem.

Key Questions

18. Agree with your group on what the denominator for the α_{HB^-} distribution function is.

19. And now what's the term in the numerator?

20. A phosphate buffer might be prepared by mixing solutions of 0.50 M NaH_2PO_4 and 0.50 M K_2HPO_4. Suppose that the pH turns out to be 6.9. Phosphoric acid has $K_{a1} = 7.11 \times 10^{-3}$, $K_{a2} = 6.32 \times 10^{-8}$, and $K_{a3} = 7.1 \times 10^{-13}$. Calculate the molar concentration of the hydrogen phosphate ion, $[HPO_4^{2-}]$, in this solution.

21. You've now seen ionic distribution functions for monoprotic, diprotic and triprotic weak acids. Discuss among yourselves and come to consensus regarding whether or not these functions might be easily extended to higher order weak acids. Be prepared to justify your conclusion.

22. Except in the case of a particular application, there's probably little utility in compiling seemingly unrelated formulae for the calculation of α-fractions pertaining to a plethora of special cases. Think about this with your group. What information do α-fractions provide? Are they simply esoteric formulae meant to amuse analytical chemists? Are α-fractions useful in general? Describe briefly how your group arrived at your conclusion.

Application

23. Calculate the molar concentration of the zwitterion of phenylalanine in a 0.0010 M solution of phenylalanine adjusted to pH 8.00. The pK_a values for phenylalanine are 2.20 and 9.31.

24. Ethylenediaminetetraacetic acid (EDTA) is a polyprotic acid generally represented as H_6Y^{2+}. The Y^{4-} form of EDTA is routinely used as a ligand for metal ion analyses. The concentration of Y^{4-} isn't very large except in quite basic solutions, so it's possible to use pH as a way to introduce selectivity into these analyses. For that reason it's frequently desirable to know $[Y^{4-}]$ in a solution at a particular pH. Look up the six pK_a values for EDTA and calculate α_Y at pH 10.00. Note that some textbooks ignore the first two pK_as because they're very strong, and only list four. If you just use the four values, your result will be indistinguishable from that obtained using all six. For the uninitiated this is a sufficiently arduous process and yet is so useful that most textbooks also contain a table of α_{Y4-} at various pH values. Check your result with such a table.

25. Disinfection of potable and recreational water is often accomplished through a process referred to as chlorination. More precisely, hypochlorite is added to the water which then results in establishment of the hypochlorous acid – hypochlorite equilibrium. Hypochlorous acid, HOCl, is the biologically active half of this equilibrium, but the usual approach to disinfection of water is to add sodium hypochlorite solution (bleach) which also contains fairly concentrated NaOH. This causes the pH of the water to creep basic with time.

a. Suppose we add enough NaOCl solution to a swimming pool to give us an initial concentration around 1.5 ppm which, according to a common theory, is ideal, but nobody's minding the store with regard to pH and it gets up around 8.5. What is the actual HOCl concentration in ppm under these conditions? Just as a reference, the minimum HOCl concentration that's considered biologically effective is 0.5 ppm.

b. Over what pH range would the HOCl concentration be biologically effective?

26. At the start of the season, Dave adds about 100 lb of baking soda ($NaHCO_3$) to the 100,000 gallon camp swimming pool. The initial concentration of sodium bicarbonate comes out to be 0.0015 M. The pH is 7.6. H_2CO_3 is a perfectly good diprotic acid as long as no CO_2 gets away, which it doesn't until its concentration gets up toward 0.1 M or the water gets really hot. For carbonic acid $K_{a1} = 4.45 \times 10^{-7}$ and $K_{a2} = 4.69 \times 10^{-11}$.

a. Calculate $\alpha_{CO_3^{2-}}$, the fraction of all the carbonate present that's in the CO_3^{2-} form under these conditions.

b. Calculate the concentration of the CO_3^{2-} ion in this water.

27. It is necessary to add Ca^{2+} to the water in plaster-lined swimming pools or the plaster will slowly dissolve. Suppose that Dave adds enough $CaCl_2$ to his pool so the hardness of the water is 200 ppm $CaCO_3$ (that is, it contains enough Ca^{2+} to make 200 ppm $CaCO_3$).

a. Calculate $[Ca^{2+}]$ in this water.

b. Are we likely to observe precipitation of $CaCO_3$ in this pool under these conditions? That is, is the K_{sp} of $CaCO_3$ exceeded?

c. Calculate how much CO_2 is dissolved in this water (that is, $[H_2CO_3]$).

28. Ethylenediamine is a highly corrosive liquid that also forms coordination complexes with many metals provided that the pH is adjusted to make a suitable concentration of the free base available. The neutral ethylenediamine molecule, ordinarily given the symbol *en*, contains two nitrogen base functions, so the fully protonated form of *en* would be written something like H_2en^{2+}.

a. Write balanced chemical equations for the equilibria involving en in water.

b. A solution initially containing 0.050 M ethylenediamine is adjusted to pH 10.0 with HCl. What is the actual molar concentration of the uncharged ethylenediamine form under these conditions?

The Buffer Zone:

What are buffers and in what pH range are they effective?

Learning Objectives

Students should be able to:

Content

- Explain how the pH of a solution is altered by the concentrations of buffer components.

- Determine the relationship between the pKa of an acid and the optimal pH range of a buffer in order to select an appropriate buffer for a particular pH range.

- Calculate the buffer capacity of a solution.

Process

- Interpret tabulated information. (Information processing)

- Recognize and predict trends in data. (Critical thinking)

- Generalize problem solutions. (Problem solving and Critical thinking)

- Include all group members. (Teamwork)

Prior knowledge

- Weak acid/base equilibrium calculations.

- Conjugate acid/base pairs identity.

Further Reading

- Harris, D.C. 2007. *Quantitative Chemical Analysis,* 7th Edition, pp.167-176. New York: W.H. Freeman

- Po, H.N., and N.M. Senozan. 2001. "The Henderson-Hasselbalch Equation: Its History and Limitations." J. Chem. Educ. 78 (11) 1499.

Authors

Mary Walczak, Caryl Fish, and Kathleen Cornely

PGLANA00XXXXa

Consider this...

Tris(hydroxymethyl)aminomethane (Tris for short) is a weak base with a $pK_a = 8.20$ at 20°C. The base can be protonated on the amino group to form the weak acid, which we will refer to as Tris acid.

Tris base

Key Questions

1. Draw the structure of Tris acid.

2. We can use a shorthand form to label a weak acid as "HA" and its corresponding conjugate base "A." The "A" and "HA" labels should be used with the appropriate charges of the species indicated.

   Label both the Tris base structure shown above and the Tris acid structure you drew in Q1 using the shorthand form described.

   Buffers contain measurable amounts of both the weak acid and the conjugate base forms. The formal concentration, F, of a buffer is defined as the sum of the concentrations of both the weak acid and the conjugate forms constituting the buffer.

3. Write an equation in which you define the formal concentration of a Tris buffer, using the notation you devised in Q2.

Table 1 includes the concentrations of both the weak acid and conjugate base forms of Tris for a series of solutions. We will fill in the blank spaces of Table 1 as we answer the Key Questions that follow.

Table 1 The pH of solutions with varying concentrations of Tris acid and Tris base			
1	2	3	4
Solution	[HA+], M	[A], M	Solution pH
1	0.005	0.095	9.48
2	0.010	0.090	9.15
3	0.020	0.080	8.80
4	0.030	0.070	
5	0.040	0.060	8.38
6	0.050	0.050	8.20
7	0.060	0.040	
8	0.070	0.030	7.83
9	0.080	0.020	
10	0.090	0.010	7.25
11	0.095	0.005	6.92

4. Divide Solutions 1-11 in Table 1 amongst the members of the group and calculate the formal concentration of each solution, using the equation you wrote in Q3. Compare answers within the group and come to a consensus.

5. **a.** Examine Columns 2 and 4. How does the pH of a solution change when the concentration of the weak acid form of Tris increases?

   b. Examine Column 3 and 4. How does the pH of a solution change when the concentration of the conjugate base form of Tris increases?

6. Which of the statement(s) below is(are) true? Circle the true statement(s).

   a. The solution with a pH of 9.48 has a formal concentration of 0.100 M.

   b. The solution with a pH of 8.38 has a formal concentration of 0.100 M.

   c. The solution with a pH of 6.92 has a formal concentration of 0.100 M.

7. Support or refute the following statement: "The formal concentration is a convenient way to refer to the concentration of a buffer."

8. What is the pH of the solution when the concentrations of the Tris acid and Tris base are both equal? Which solution in Table 1 fits this description?

9. The data in Table 1 can be plotted in which the pH is graphed as a function of concentration of the buffer components. The plot is shown in Figure 1 below.

   a. Which curve corresponds to the HA^+ concentration? Label the plot.

   b. Which curve corresponds to the A concentration? Label the plot.

   c. At what point does $pH = pK_a$? Label the plot.

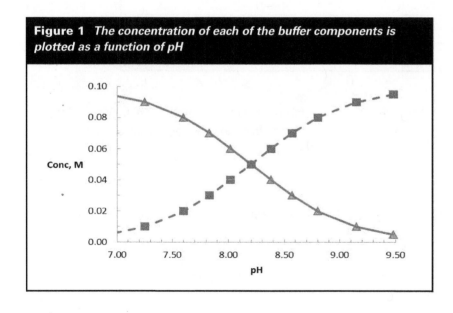

Figure 1 *The concentration of each of the buffer components is plotted as a function of pH*

10. Using Figure 1, answer the following questions:

   a. What happens to the pH (relative to the pK_a) when the concentration of Tris acid is greater than 0.050 M?

   b. What happens to the pH (relative to the pK_a) when the concentration of Tris acid is less than 0.050 M?

   c. What happens to the pH (relative to the pK_a) when the concentration of Tris acid is equal to 0.050 M?

A **buffered solution** (a "buffer") resists changes in pH when strong acids or strong bases are added or when dilution occurs.

A solution contains a weak acid, HA, and its conjugate base A⁻. In many cases, the pH of a buffer can be calculated by using the Henderson-Hasselbalch equation:

$$pH = pK_a + log\left(\frac{[A^-]}{[HA]}\right)$$

Note that the Henderson-Hasselbalch equation is derived on the basis of some significant assumptions. Chief among these assumptions is ignoring any dissociation of HA or hydrolysis of A⁻. In other words, the concentrations of HA and A⁻ are assumed to be equal to their formal concentrations. The dissociation of water is also ignored. These approximations are valid for acids with pK_a values in the range of ~5-9, for concentrations greater than ~1 mM or less than ~1 M, or for situations in which the two components (HA or A⁻) are close to the same concentration (i.e., within two orders of magnitude.)

Also note that the familiar form of the Henderson-Hasselbalch equation is presented in which the weak acid form is expressed as HA and the conjugate base form is expressed as A⁻, but for Tris the buffer components should be expressed as HA⁺ and A, respectively, as noted in Q2.

11. Assign each of the three empty pH cells in Column 4 of Table 1 to group members. Use the Henderson-Hasselbalch equation to calculate the pH values for these solutions. Have at least two students calculate each value to check for accuracy. Compare your answers and agree upon the correct values before proceeding.

12. Do your calculated values follow the trends in Table 1 as described in Q5?

Consider this...

The pH of solutions with varying concentrations of Tris acid and Tris base, first introduced in Table 1, are shown below in Table 2. In addition, the solution pH after the addition of either strong acid or strong base to these solutions is shown. In general, when strong acid or base is added to a solution, we expect that the pH of the solution will change. To a small extent, this occurs in buffer solutions (Solutions 1-11) and to a much larger extent in an unbuffered solution (Solution 12). The resulting pH values of the original solution *after either* strong acid (HCl) *or* strong base (NaOH) is added are shown in Columns 5 and 7 in Table 2 below. We will complete Table 2 as we progress through the Key Questions below.

Table 2 *The solution pH after the addition of 0.00050 moles of either HCl or NaOH to 100.0 mL of buffer is indicated. Note the addition of Solution 12 to the table, which does not contain Tris.*

1	2	3	4	5	6	7	8
Solution	[HA$^+$], M	[A$^-$], M	Solution pH	pH after addition of 0.00050 moles HCl	pH change	pH after addition of 0.00050 moles NaOH	pH change
1	0.005	0.095	9.48	9.15		10.60	
2	0.010	0.090	9.15	8.95		9.48	
3	0.020	0.080	8.80	8.68		8.95	
4	0.030	0.070	8.57	8.47		8.68	
5	0.040	0.060	8.38				
6	0.050	0.050	8.20	8.11		8.29	
7	0.060	0.040	8.02	7.93		8.11	
8	0.070	0.030	7.83	7.72		7.93	
9	0.080	0.020	7.60	7.45		7.72	
10	0.090	0.010	7.25	6.92		7.45	
11	0.095	0.005	6.92	4.60			
12	0.000	0.000	7.00				

13. Consider Solution 12 in Table 2.

    a. Write a chemical reaction that describes what happens when HCl is added to Solution 12.

    b. Write a chemical reaction that describes what happens when NaOH is added to Solution 12.

14. Divide the group in half; one subgroup completes part a and the other completes part b.

    a. Calculate the pH of the resulting solution after the addition of 0.00050 moles of HCl to 100.0 mL of Solution 12. (Assume a negligible volume change.) Add your Solution 12 value to Table 1.

    b. Calculate the pH of the resulting solution after the addition of 0.00050 moles of NaOH to 100.0 mL of Solution 12. (Assume a negligible volume change.) Add your Solution 12 value to Table 1.

15. a. Compare Columns 4 and 5 for Solution 12, refer to the equation you wrote in Q13a, then read the statements below. Circle the statement that explains the reason for the observed pH change:

    (i) The solution pH decreases because the $[H^+]$ decreases when HCl is added.

    (ii) The solution pH increases because the $[H^+]$ decreases when HCl is added.

    (iii) The solution pH decreases because the $[H^+]$ increases when HCl is added.

    (iv) The solution pH increases because the $[H^+]$ increases when HCl is added.

    b. Compare Columns 4 and 7 for Solution 12, refer to the equation you wrote in Q13b, then read the statements below. Circle the statement that explains the reason for the observed pH change:

    (i) The solution pH decreases because the $[OH^-]$ decreases when NaOH is added.

    (ii) The solution pH increases because the $[OH^-]$ decreases when NaOH is added.

    (iii) The solution pH decreases because the $[OH^-]$ increases when NaOH is added.

    (iv) The solution pH increases because the $[OH^-]$ increases when NaOH is added.

We will now consider **Solutions 1-11** in Table 2.

16. Write a balanced chemical reaction (including structural formulas) that describes the dissociation of Tris acid in water.

17. Table 2 is missing three pH values for the addition of HCl and NaOH in Columns 5 and 7. As a group, design two equations that you can use to determine each of the missing pH values—one equation for added acid, a second equation for added base. When designing your equations, consider that the added HCl or NaOH will react with either the Tris acid or Tris base in solution. Come to a consensus on your strategy, and then compare your strategy to that of another group. Refine your group's strategy if necessary before proceeding.

Using the equations you have written, and assuming an initial volume of 100.0 mL (and that volume changes upon addition of HCl or NaOH are negligible), carry out the calculation and add the missing values to Table 2 in Columns 5 and 7. Do your results follow the trends for those columns?

18. Calculate the pH changes for the set of Solutions 1-12 when HCl is added and when NaOH is added. Add these values to Table 2 in Columns 6 and 8.

19. Compare the pH changes that occur when HCl or NaOH is added to Solution 12 with the pH changes that occur in Solutions 1-11, then circle the statement below that best explains the observed changes.

"When HCl or NaOH is added to a solution, the pH change is greatest...

(i) ...when the [Tris acid] is greatest."

(ii) ...when the [Tris base] is greatest."

(iii) ...when the concentrations of [Tris acid] and [Tris base] are about equal."

(iv) ...when both Tris base and Tris acid are absent."

20. Compare the pH changes that occur when HCl or NaOH is added to Solution 12 with the pH changes that occur in Solutions 1-11, then circle the statement below that best explains the observed changes.

"When HCl or NaOH is added to a solution, the pH change is the smallest...

(i) ...when the [Tris acid] is greatest."

(ii) ...when the [Tris base] is greatest."

(iii) ...when the concentrations of [Tris acid] and [Tris base] are about equal."

(iv) ...when both Tris base and Tris acid are absent."

21. Effective buffers resist changes in pH when strong acid or strong base is added. Answer the following questions regarding the effectiveness of Tris as a buffer:

a. Is Tris a more effective buffer at pH = 7 or pH = 8?

b. Is Tris a more effective buffer at pH = 7 or pH = 9?

22. Use the equations you wrote in Q13 and Q16 and LeChatelier's principle to describe the equilibrium changes that occur when either HCl or NaOH is added to a solution containing both Tris acid and Tris base. Explain how these equilibrium changes allow the solution to resist changes in pH when either the strong acid or strong base is added.

23. Which solution(s) is (are) most resistant to changes in pH? Explain why solutions with similar concentrations of HA^+ and A are least sensitive to the addition of strong acid and base.

Consider this...

In this portion of the activity, we will focus on the concept of effective buffering range. At pH values that encompass the effective range, sufficient amounts of both the weak acid and the conjugate base are present. Outside of the buffering range, the amount of one of the species, either the weak acid or the conjugate base, is present in sufficiently small concentrations that buffering is no longer effective.

Enzymes are biological catalysts that require a specific solution pH in order to function effectively. Figure 2 shows pH-activity curves for enzymes A and B.

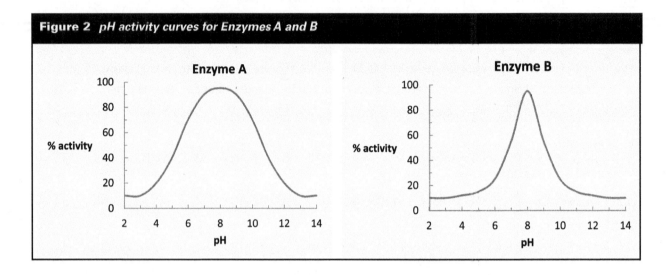

Figure 2 *pH activity curves for Enzymes A and B*

24. Use the data presented in Figure 2 to answer the following questions:

a. What is the optimal pH for Enzyme A? For Enzyme B?

b. Over what pH range is Enzyme A at least 90% active? Enzyme B?

25. a. An experiment is carried out using Enzyme A. During the course of the experiment, either H^+ or OH^- ions are released by other solution components. Which of the solutions presented in Table 2 could be used to dissolve Enzyme A for this experiment if it is essential that the enzyme be at least 90% active?

b. Which of the solutions presented in Table 2 could be used to dissolve the enzyme for a similar experiment involving Enzyme B in which it is essential that the enzyme be at least 90% active?

> An effective buffering range is expressed as an interval that includes the pK_a of the buffer, i.e., $pK_a \pm$ ___.

26. How would you specify the effective buffering range for Tris buffer for the experiment described in Q25 involving Enzyme A?

Would you express the effective buffering range differently for the experiment described in Q25 involving Enzyme B?

27. In 3-4 grammatically correct sentences, summarize the important principles that your group has learned about buffers in the course of completing this activity.

Consider this...

Buffer capacity is defined as the ability of a solution to resist changes in pH when small amounts of strong acid or base are added. Buffer capacity can be mathematically defined as the negative of the first derivative of the amount of acid added with respect to pH, or $\dfrac{d[H^+]}{dpH}$

Using this definition, the Henderson-Hasselbalch equation, and the 2.303 term to convert from ln to $\log_{10}$, we can derive the following expression for buffer capacity:

$$Buffer\ capacity = 2.303 \times \frac{[A^-][HA]}{[A^-]+[HA]}$$

28. The [Tris acid], [Tris base] and pH for the solutions from Table 1 are shown in Table 3 below. Calculate the buffer capacities for Solutions 4, 6, 9 and 10. Divide the work amongst group members and have at least one group member check another's calculations. Place your answers in Table 4 below, and agree upon your values before proceeding.

Table 3 *Buffer capacities of solutions with varying concentrations of Tris acid and Tris base*

Solution	[Tris acid], M	[Tris base], M	pH	Buffer capacity
1	0.005	0.095	9.48	0.011
2	0.010	0.090	9.15	0.021
3	0.020	0.080	8.80	0.037
4	0.030	0.070	8.57	
5	0.040	0.060	8.38	0.055
6	0.050	0.050	8.20	
7	0.060	0.040	8.02	0.055
8	0.070	0.030	7.83	0.048
9	0.080	0.020	7.60	
10	0.090	0.010	7.25	
11	0.095	0.005	6.92	0.011

29. Which solution(s) shown in Table 3 has(have) the greatest buffer capacity? The least?

30. Write a sentence that describes the conditions of [Tris acid] and [Tris base] that exist in the solution with the highest buffer capacity. Write a second sentence that describes the relationship between pH and pK_a in the solution with the highest buffer capacity.

31. Compare Solution 6 with another solution that we will call Solution 13. Write the value of the buffer capacity for Solution 6 that you already calculated in Table 4 below. Calculate the buffer capacity for Solution 13 and place its value in Table 4 as well. Fill in the values for the [Tris acid]:[Tris base] ratio and the formal concentrations for the two solutions.

Table 4 Buffer capacities of two buffer solutions with different formal concentrations

Solution	[Tris acid], M	[Tris base], M	$\dfrac{\text{[Tris base]}}{\text{[Tris acid]}}$	F	Buffer capacity
6	0.050	0.050			
13	0.500	0.500			

32. Which solution has the greater buffer capacity, Solution 6 or Solution 13?

33. a. What is the magnitude of the difference in [Tris acid] and [Tris base] between Solutions 6 and 13?

b. What is the magnitude of the difference in buffer capacity between Solutions 6 and 13?

34. Circle the statement below that is true.

At a fixed ratio of [HA$^+$] : [A]....

...the buffer capacity decreases when the formal concentration increases.

...the buffer capacity increases when the formal concentration increases.

...the buffer capacity stays the same when the formal concentration increases.

35. Summarize what you have learned in this activity by writing 1-2 grammatically correct sentences that describe the two most important factors to consider when evaluating buffer effectiveness.

36. List at least three concepts you learned about buffers as a result of completing this activity.

37. Are there any concepts that any member of the team still finds confusing?

38. How confident do you feel about your ability to assess buffer effectiveness?

Application

The weak acid and conjugate base concentrations for three different solutions are shown in Table 5. Each solution has a formal concentration of 0.10 M.

Table 5 *The weak acid (HA) concentrations for selected weak acids at various pH values*

	pK_a = 7.55		pK_a = 6.35		pK_a = 6.86	
	HEPES (4-(2-hydroxyethyl)-1-piperazine-ethanesulfonic acid)		Carbonic acid, H_2CO_3		Phosphate, monobasic, $H_2PO_4^-$	
pH	$[HA^+]$, M	$[A]$, M	$[HA]$, M	$[A^-]$, M	$[H_2A^-]$, M	$[HA^{2-}]$, M
1.0	0.10	0.00	0.10	0.00		
1.5	0.10	0.00	0.10		H_3A predominates	The $[H_2A^-]$ is negligible.
2.0	0.10		0.10	0.00		
2.5	0.10	0.00	0.10			
3.0	0.10		0.10		0.09	0.00
3.5	0.10	0.00	0.10	0.00	0.10	0.00
4.0	0.10		0.10		0.10	0.00
4.5	0.10	0.00	0.10		0.10	
5.0	0.10		0.10	0.00	0.10	0.00
5.5	0.10		0.09		0.10	
6.0	0.10		0.07		0.09	0.01
6.5	0.09		0.04	0.06	0.08	
7.0	0.08	0.02	0.02		0.06	
7.5	0.05		0.01	0.09	0.03	
8.0	0.03		0.00	0.10	0.01	
9.0	0.00	0.10	0.00		0.00	

39. Use Table 5 to answer the following application questions:

    a. Calculate the [A⁻] concentration (to two decimal places) at each pH. Display the values in Table 5.

    b. Use the data in Table 5 to determine the effective buffering ranges for the weak acids listed. Display your answers in Table 6 below. Each group member should answer this question individually, then compare your answers amongst group members.

Table 6 *Effective buffering ranges for selected weak acids*

Weak acid	pK_a	Effective buffering range
HEPES		
Carbonic acid		
Monobasic phosphate		

    c. You are running a kinetics experiment that requires that the pH be held at 6.5 ± 0.5. Which of the buffer(s) listed in Table 6 could be used for this experiment? Which buffer(s) could not be used?

40. In order to properly function, enzymes and other proteins must retain their three-dimensional structures. This structure is pH-dependent because many of the amino acid side chains' -R groups contain acidic or basic functional groups. In the blood, the pH is maintained by a complex variety of buffers, including the carbonic acid/bicarbonate buffer. Given that blood pH is maintained within a narrow range of 7.35-7.45, evaluate nature's choice of this particular buffer system.

41. Carbonic acid, H_2CO_3, and bicarbonate, HCO_3^- are present in the blood and play a major role in regulating blood pH, which is maintained within a narrow range of 7.35 to 7.45 in a healthy person. Acidosis results if the pH level drops below 7.35; alkalosis if the pH is raised above 7.45. A patient suffers from alkalosis as a result of an overdose of aspirin tablets. In the emergency room, the physician measures her blood pH to be equal to 7.55. Determine the ratio of HCO_3^- to H_2CO_3 in the patient's blood at this pH. How does this ratio compare to the ratio of HCO_3^- to H_2CO_3 in normal blood? (Use an average value of 7.40 for the pH of normal blood.) Can the H_2CO_3/HCO_3^- system work effectively as a buffer in this patient under these conditions? Explain, using concepts developed in this activity.

42. The patient described in Q41 has a blood bicarbonate concentration of 18 mM two hours after being brought to the emergency room. Normal blood bicarbonate concentrations are about 24 mM. Calculate the concentration of H_2CO_3 in the patient's blood and compare this value to the concentration of H_2CO_3 found in a normal person's blood. Can the H_2CO_3/HCO_3^- system work effectively as a buffer in this patient under these conditions? Explain, using concepts developed in this activity.

43. Metabolic acidosis is a general term that describes a number of disorders in metabolism in the body that result in a lowering of the blood pH from 7.4 to 7.35 or below. Metabolic alkalosis occurs when the blood pH rises to 7.45 or greater. The kidney plays a vital role in regulating blood pH. The kidney can either excrete or reabsorb various ions, including phosphate, $H_2PO_4^-$; ammonium, NH_4^+; or bicarbonate, HCO_3^-.

a. Which ions are excreted and which ions are reabsorbed in metabolic acidosis?

b. Which ions are excreted and which ions are reabsorbed in metabolic alkalosis? Explain, using relevant chemical equations and concepts developed in this activity.

When Acids and Bases React: Laboratory
What does a titration curve tell you about an acid/base reaction?

Learning Objectives

Students should be able to:

Content

- For a given acid/base reaction, distinguish between equivalence point and endpoint.

- Given a titration curve, identify whether a monoprotic acid is "strong" or "weak".

- Given a titration curve, identify the main features of the curve, i.e. (1) whether an analyte is an acid or base, (2) whether an analyte is a mono- or polyprotic acid, and (3) what is/are the analyte's pK value(s).

Process

- Obtain titration curve using pH probe, drop counter, and data acquisition software (Experimental technique)

- Determine the equivalence point of a titration curve from 1st and 2nd derivative plots of the curve using computer software, i.e. *Excel, Vernier* software, etc. (Information processing)

- Predict the appearance of a titration curve (critical thinking)

Prior knowledge

- Definition of pH

- Differences between strong vs. weak acid/bases

- Definition of mono- and polyprotic acids/bases

- How to perform titrations using acid/base indicator

Authors

Ruth Riter (Shepherd) and Mary Walczak

Acknowledgments

Mary Walczak's Fall 2009 Quantitative Analysis class at St. Olaf College.

PGLANA004021a

Part 1: Standardize a NaOH solution using an acid/base indicator.

Consider this...

In the first part of this experiment the concentration of a sodium hydroxide solution will be determined by standardizing it against the primary standard potassium hydrogen phthalate (KHP). Students will prepare three separate samples of KHP and titrate each with the NaOH solution to a phenolphthalein endpoint. The Data Table generated will be analyzed in the Key Questions below.

General Laboratory Procedure

A. Prepare 3 KHP samples to be titrated with the NaOH solution. Calculate the number of grams of KHP that will react with ~25 mL of the NaOH solution provided. The NaOH solution concentration will be between 0.050–0.10 M. Weigh three KHP samples of appropriate mass into three separate, clean Erlenmeyer flasks or beakers. Dissolve the KHP in ~50 mL of deionized water. Add 2-3 drops of your acid/base indicator.

B. Rinse and fill your buret with the NaOH solution. Record the concentration of the solution, if provided. Titrate the first KHP sample to the endpoint. Calculate the volume required to reach the endpoint. Titrate the other two KHP samples.

C. Prepare one beaker or Erlenmeyer flask with 50 mL of water to serve as a blank solution. Add 2-3 drops of your indicator, record the solution color and titrate with NaOH until the endpoint is reached. Calculate the volume required to titrate the blank solution.

D. Create a table of data from your four titrations. For each titration, be sure to include the initial, final and total volume of NaOH solution, the mass of KHP, the indicator used and solution observations.

Use the table as you generated in the above laboratory experiment to address the following:

Key Questions

1. According to your data, how many mL of NaOH are required for the blank solution to reach the endpoint?

2. In the titration of one of the KHP samples, does the volume required to reach the endpoint *overestimate, underestimate or correctly estimate* the volume of NaOH required to react with all the KHP in the solution?

3. Explain how the volume required to reach the endpoint in the KHP titrations should be adjusted as a result of the blank solution. Make this correct to the three KHP trials.

4. Write the balanced full ionic and net ionic chemical reaction that takes place when a KHP sample is titrated with NaOH.

5. Write a general mathematical equation to calculate the molarity of the NaOH from the mass of KHP and volume of NaOH used in a titration.

6. Divide your three KHP titrations among group members. For each solution, calculate the molarity of the NaOH solution using the equation from Q5. Have at least two students check each result.

7. Report the average molarity of the NaOH solution and the associated uncertainty. Use any statistical tests that you feel are appropriate.

8. Compare your [NaOH] from Q7 with other groups in the class. Are the answers the same? Use any statistical tests that you feel are appropriate to make this comparison.

The color change observed during a titration marks the **endpoint** of the reaction. This occurs when the acid/base reaction is complete and a small amount of the indicator reacts with the titrant. The observed color change is due to the reacted indicator species. The **equivalence point** of a titration is the point where the acid/base reaction is completed based on the stoichiometry of the neutralization reaction.

---

9. Based on your experimental data, is the point at which you observed the color change in the acid/base titrations the endpoint or the equivalence point of the reaction? Explain your reasoning.

10. How does the identity of the acid/base indicator you are using effect your [NaOH]? Explain your reasoning.

Part 2: Potentiometric Titration of an Assigned Acid Using Data Acquisition Software.

Consider this...

Now that the concentration of the NaOH solution has been determined, you will use that solution to titrate an assigned acid. You will determine the equivalence point of the titration using three graphical methods and compare these values with an endpoint obtained by titration with an indicator dye.

General Laboratory Procedure

E. Prepare a stock solution of your acid that is ~0.05 M.

F. Measure out *two* 25.00-mL aliquots of your acid and place each in an Erlenmeyer flask or beaker. (*Note:* The opening of at least one of the flasks or beakers must be large enough to put in a pH probe and add titrant.).

G. Rinse and fill a buret with your standardized NaOH solution.

H. Using pH probe and drop counter, acquire a titration curve for one of the aliquots until the pH of the solution remains constant for 5 mL of titrant. Record initial and final buret volumes and all observations. Save and print the titration curve.

I. Add two to three drops of indicator to the *remaining* aliquot and titrate it against your standardized NaOH solution using the procedure in Part 1. Determine the volume (ΔV) of NaOH needed for this titration.

Data Analysis

J. Using the data acquisition software, *Excel*, or other graphing program, plot the first and second derivative for the titration curve. Save and print each plot.

K. Determine the volume of NaOH solution at the inflection point of the titration curve, at the peak of the curve in the first derivative plot, and at the point that the second derivative passes through $y=0$.

L. Make a table summarizing the equivalence points as determined by these three different graphical methods. Add the volume of NaOH required to reach the end point in the last titration in which indicator was used.

Key Questions

11. Compare the volumes of NaOH determined from your titration curve and the corresponding 1st and 2nd derivative plots. How similar are these volumes?

12. If volumes are not the same, which volume is the "better" answer? Explain your reasoning.

13. Compare the volume of NaOH determined from your titration curve and from your titration with an indicator. How similar are they? Which is the better measure of the equivalence point? Explain your reasoning.

Analysis of Class Titration Curves

Share the titration curve of your assigned acid with every group in class and obtain copies of their curves. You need not provide your derivative plots to other groups. Use these curves to answer the following questions.

---

14. For each of the titration curves, identify the acid being titrated. Classify each acid as a strong or weak acid and as a mono-, di- or triprotic acid. Write the titration reaction(s) for each titration.

15. Examine each of the titration curves. Make a list of characteristics that are the same among two or more curves.

16. Make a list of characteristics that are different among the curves.

17. For each characteristic listed in Q15 and Q16, explain why that feature is present for the particular acid titrated.

18. What is the pH at the equivalence point of a strong acid/strong base titration?

---

19. How does the pH at the equivalence point of a weak acid/strong base titration compare to a strong acid/ strong base titration?

20. What is the pH at the equivalence point of a polyprotic acid and strong base titration? Is there more than one equivalence point? Explain your answer.

21. Examine the titration curves for the reactions of a monoprotic, strong acid and a monoprotic, weak acid with NaOH.

   a. How do the pHs of the equivalence points compare?

   b. How do the shapes of the titration curves near the equivalence compare?

   c. Describe any other observed differences in the titration curves.

   d. Describe how you would know if the analyte was a monoprotic, strong acid or a monoprotic, weak acid by looking at the titration curve.

22. After comparing the titration curves for the reactions of monoprotic and diprotic acids with NaOH, describe how the shapes of the titration curves compare. In other words, given a titration curve, describe how you would know if the analyte was a monoprotic or diprotic acid.

23. Using your textbook or other reference, look up the pK_a values for each of the weak acids. Using the titration curves for two or more monprotic weak acid, mark the location on the curve where $pH=pK_a$. Is there any common feature(s) about this point on all of the graphs? Explain your reasoning.

24. Describe in your own words a procedure for determining the experimental pK_a of weak acid from a titration curve.

Application

25. You are given a solution of diethylamine $((CH_3CH_2)_2NH, pK_b=3.23)$ to titrate with a solution of HCl.

    A. Write a balanced chemical reaction for this reaction.

    B. Which solution is the analyte?

    C. Sketch a titration curve of pH vs. volume of titrant.

    D. Sketch the first and second derivatives of the titration curve.

    E. Identify the equivalence point and the point where $pH=pK_a$.

26. Given a titration curve:

    A. Identify equivalency point(s).

    B. Identify analyte (acid/base, weak/strong, mono/polyprotic) and explain your reasoning.

    C. If weak acid or base, determine the pK value(s).

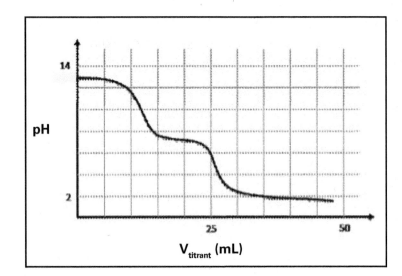

Electrochemistry: The Microscopic View of Electrochemistry

Learning Objectives

Students should be able to:

Content

- Describe microscopic-level processes occurring at electrodes and explain how these processes result in formation of the electrochemical double layer.
- Explain the origins of electrochemical potential.
- Use E° values to predict the spontaneous direction in electrochemical reactions.

Process

- Interpret pictures of electrochemical processes. (Information Processing)
- Draw and interpret microscopic pictures of electrochemical processes. (Information Processing)
- Draw and interpret graphical representations of electrochemical processes. (Critical Thinking)

Prior knowledge

- General understanding of electrostatics (e.g., charged particle behavior).
- Definitions of oxidation, reduction and standard reduction potentials.
- The use of ΔG as a predictor of spontaneous reactions.

Further Reading

- D.C. Harris, *Quantitative Chemical Analysis,* 7th Edition, 2007 W.H. Freeman: USA, Sections 14-1 through 14-3, pp. 270-9.
- D.A. Skoog, D.M. West, F.J. Holler, S.R. Crouch, Fundamentals of Analytical Chemistry, 8th Edition, 2004 Thompson Brooks/Cole: USA, Sections 18B, 18C-1 through 18C-3, pp. 496-508.
- Thomas Greenbowe's electrochemistry animation site: www.chem.iastate.edu/group/Greenbowe/sections/projectfolder/animationsindex.htm
- Özkaya, A.R., Üce, M. and Sahin, M., J. *Chem. Educ.* 2006 83(11) pp.1719-23.
- Electrochemistry Dictionary: www.electrochem.cwru.edu/ed/dict.htm
- Martins, G.F. Why the Daniell Cell Works! *J. Chem. Educ.,* 1990, 67 (6), pp. 482.

Authors

Christine Dalton and Mary Walczak

PGLANA004022a

Section 1: The Microscopic View of Electrochemistry

Consider this...

Figure 1(a) below contains a strip of zinc metal and a beaker containing a solution of the corresponding metal ion (e.g., $Zn(NO_3)_2(aq)$).

In Figure 1(b), the Zn metal strip is placed in the Zn^{2+} solution. The (+) and (-) signs are used to represent the electrostatic interface that forms at the metal/solution interface. These signs do not indicate the positive and negative ions in solution or free electrons in the metal electrode, but they represent the overall charge on the metal electrode and the overall charge of the solution near the metal electrode. The signs in Figure 1(b) are not meant to be absolute values, but rather relative to one another.

Figure 1(c) is the microscopic view of the interface between the surface of the electrode and the metal ion solution within which the metal electrode is placed, not showing NO_3^- and H_2O for simplicity. Electrons within the metal surface are delocalized.

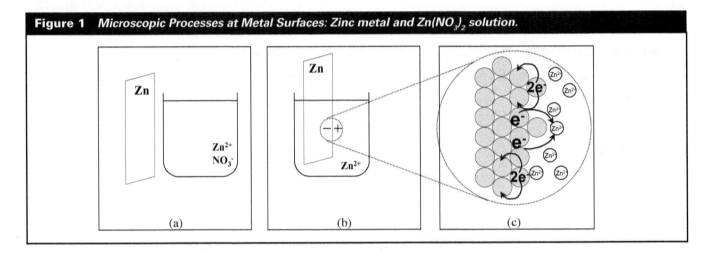

Figure 1 *Microscopic Processes at Metal Surfaces: Zinc metal and $Zn(NO_3)_2$ solution.*

Key Questions

1. Examine the three panels in Figure 1. Classify each panel as representative of the situation *before* or *after* immersion.

    (a)   (b)   (c)

    What is the relationship between panels (b) and (c)?

2. A Zn^{2+} ion collides with the electrode and gains 2 e-. What is the reaction for this event? Is this an oxidation or a reduction?

---

3. A Zn atom on the electrode loses 2 e- which join the collection of delocalized electrons on the metal. What is the reaction for this event? Is this an oxidation or a reduction?

4. Examine figure 1(c) closely. Three microscopic processes are shown. One is characterized by movement of delocalized electrons to a zinc ion. Two others show the loss of electrons by a Zn atom. Label the oxidation and reduction processes in figure 1c.

Key Questions

A strip of zinc metal is placed in a solution of $Zn(NO_3)_2$ and equilibrium is established. Note: Nitrate ions and water molecules are omitted for clarity.

Figure 2 *Predominant Microscopic Processes at Metal Surfaces: Zinc metal in $Zn(NO_3)_2$ solution (a) when initially immersed and (b) once equilibrium is reached.*

(a) Before equilibrium   (b) At equilibrium

---

5. Draw arrows to show where the electrons are going as equilibrium is reached. Confirm that the drawing shows that *oxidation* is the favored reaction.

6. Describe the microscopic processes that occur when a strip of zinc metal is placed in a solution of $Zn(NO_3)_2$.

7. Which of the following reactions occurs to a greater extent? Explain your reasoning.

$$Zn^{2+}(aq) + 2e- \rightarrow Zn(s) \qquad Zn(s) \rightarrow Zn^{2+}(aq) + 2e-$$

8. Divide the following two questions among group members. Circle your answer to the following questions. Come to consensus on your answers.

Based on your answer to Q5-Q7, is the charge at the surface of the Zn electrode:

Positive   Negative   Neutral?

Explain your reasoning.

Based on your answer to Q5-Q7, is the charge in the solution in the vicinity near the electrode:

Positive   Negative   Neutral?

Explain your reasoning.

9. Verify your answers to Q8 by looking at Figure 1b.

Consider this...

The following figure shows the Zn electrode immersed in a $Zn(NO_3)_2$ solution at equilibrium. Note: Solvent molecules have been removed for simplicity. Suppose you have a tiny charge measuring device that you can position at various locations in the solution to measure the charge at a single point.

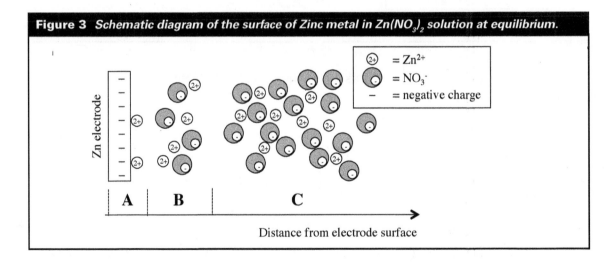

Figure 3 *Schematic diagram of the surface of Zinc metal in $Zn(NO_3)_2$ solution at equilibrium.*

10. What is the overall charge in each of the following regions of the figure?

Region A	Positive	Neutral	Negative
Region B	Positive	Neutral	Negative
Region C	Positive	Neutral	Negative
Regions A and B	Positive	Neutral	Negative
Regions B and C	Positive	Neutral	Negative

11. How does the charge measured by your device change if the probe is moved from point A to point B in the figure?

12. What happens to the local charge when you move the probe from point B to point C?

13. On the following graph of charge versus distance from the electrode, label the location of points A, B, and C from Figure 3. As a group, use the graph to explain what happens to the charge as the distance from the electrode increases.

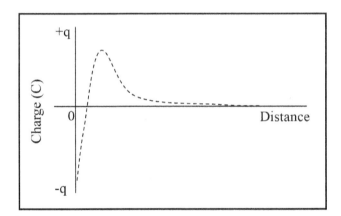

The electrical double layer (EDL) occurs at the interface between an electrode and the solution. It is comprised of adjacent positive and negative regions, as shown in Figure 3. The inner layer (A) has charged particles directly adsorbed onto the surface of the electrode, while the outer layer (B) has ions electrostatically attracted to the adsorbed charged particles. This second layer is called the diffuse layer. The net electric charge in the diffuse layer is equal in magnitude to the net electric charge on the surface but with the opposite charge, leading to a net electrically neutral EDL.

---

14. Indicate the location of the EDL on Figure 3 and on the graph in Q13.

Consider this...

Figure 4(a) below contains a strip of lead metal and a beaker containing a solution of the corresponding metal ion (e.g., $Pb(NO_3)_2 (aq)$).

In Figure 4(b), the Pb strip is placed in the Pb^{2+} solution.

Figure 4c is the microscopic view of the interface between the surface of the electrode and the metal ion solution within which the metal electrode is placed, not showing NO_{3-} and H_2O for simplicity.

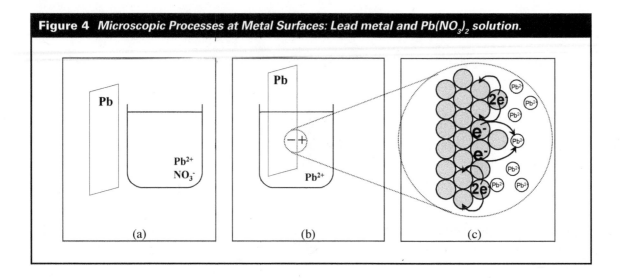

Figure 4 *Microscopic Processes at Metal Surfaces: Lead metal and $Pb(NO_3)_2$ solution.*

Key Questions

15. Divide into pairs to answer the following question: According to figure 4(c) which of the following reactions occur immediately after a strip of lead metal is placed into a solution of $Pb(NO_3)_2 (aq)$?

   I. $Pb^{2+}(aq) + 2e- \rightarrow Pb(s)$
   II. $Pb(s) \rightarrow Pb^{2+}(aq) + 2e-$

   A) I only   B) II only   C) I and II   D) none

Share your answer with the other pair. Come to consensus and explain the rationale behind your choice.

16. Using the figures below as a starting point, draw the microscopic processes that occur when a strip of lead metal is placed in a solution of $Pb(NO_3)_2$. Draw in the arrows to show where the electrons are going. Note: Nitrate is omitted for clarity.

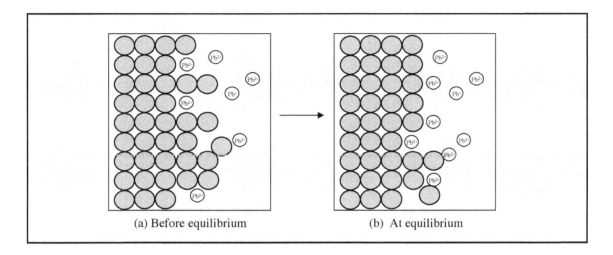

(a) Before equilibrium   (b) At equilibrium

17. Now that you have explored a Zn/Zn^{2+} and Pb/Pb^{2+} metal-solution interface, list the differences you have discovered between the two interfaces.

18. Using Figure 3 as a model, draw the charge distribution (electrical double layer) AT the surface of the electrode (point A), farther away from the electrode (point B), and far away from the electrode (point C) for the Pb/Pb^{2+} system. Label the inner layer, the diffuse (outer) layer and the bulk solution. Make sure your drawing shows the differences you identified in Q17.

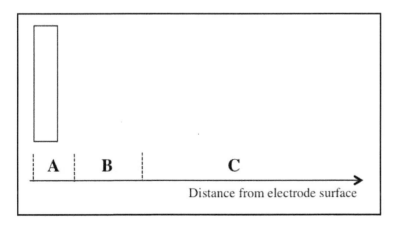

A   B   C

Distance from electrode surface

19. Compare Zn and Pb in the following graph of charge versus distance from the electrode.

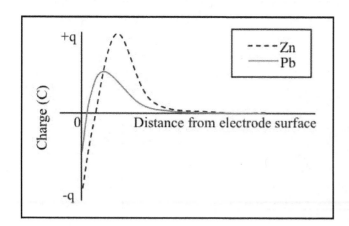

(a) Which electrode has a double layer that extends farther into solution from the metal surface?

(b) Which electrode has the larger residual negative charge at the metal surface?

(c) What evidence from the graph led to your answers in (a) and (b)?

20. The following drawings of double layer region represent the charge distribution for a Zn and for a Pb electrode immersed in a solution of its nitrate salt. Identify which picture represents Zn and which picture illustrates Pb. List two criteria you used in arriving at your answer.

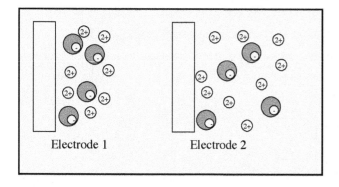

21. Using the information in the previous two questions, list the similarities and differences in the charge distribution for Zn and Pb electrodes in solution. Compare your answers with another group.

22. In your own words, define the electrical double layer for a metal placed into a solution containing the metal cation (i.e., Zn/Zn^{2+} or Pb/Pb^{2+}).

Section 2: The Energetics of Oxidation-Reduction Reactions

Consider this...

The following energy level diagrams show a three-step process for converting a metal atom into a hydrated divalent cation.

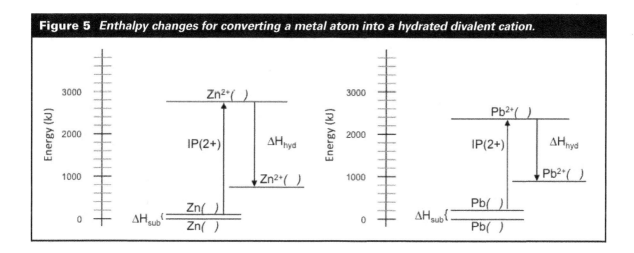

Figure 5 *Enthalpy changes for converting a metal atom into a hydrated divalent cation.*

Enthalpy is a state function. The change in enthalpy between the initial and final species is the same, regardless of path. As such, the path from one chemical species to another does not matter.

---

23. The first step of the process, annotated as ΔH_{sub} in the figure, is the sublimation of zinc atoms. The value of ΔH_{sub} is 130 kJ for Zn and 197 kJ for Pb. On the figure, write in the states of matter for the Zn in the sublimation of zinc. Write the chemical reaction that occurs for the sublimation step.

24. The second step of the process, annotated as IP(2+) in the figure, is the ionization of the gas phase atom to the divalent cation. The value of IP(2+) is 2641 kJ for Zn and 2168 kJ for Pb. On the figure, write in the state of matter for the zinc divalent cation after gas phase ionization. Write the chemical reaction that occurs during the gas phase ionization of zinc atoms.

25. The third step of the process, annotated as ΔH_{hyd} in the figure, is hydration of the divalent cation. The value of ΔH_{hyd} is -2009 kJ for Zn and -1486 kJ for Pb. On the figure, write in the state of matter for the divalent cation after hydration. Write the chemical reaction that occurs during the hydration of divalent zinc cations.

26. Write the overall chemical reaction for Zn that is represented in the above figure. Verify that you have the correct reaction by comparing your answer to the figure legend.

27. Repeat Q23-Q26 for Pb.

28. The following table summarizes the enthalpy of sublimation, the ionization potential, and the enthalpy of hydration for Zn and Pb. Calculate enthalpy change of reaction for Zn and Pb and add your answers to the table. Compare the enthalpy of reaction that you calculated for Zn and Pb.

Metal	ΔH_{sub} (kJ)	IE (2+) (kJ)	ΔH_{hyd} (kJ)	ΔH_{rxn} (kJ)
Zn	130	2641	−2009	
Pb	197	2168	−1486	

29. Based on your answer to Q19 about the electrical double layer, which metal is easier to oxidize, Zn or Pb?

30. Do the relative magnitudes of the ΔH_{rxn} values calculated in Q28 support or refute the conclusions from the trend from Q19? Explain your reasoning.

31. A piece of Zn metal is placed into a solution of $Pb(NO_3)_2$. The predicted reactions are:

$Zn(s) \rightarrow Zn^{2+}(aq) + 2e-$

$Pb^{2+}(aq) + 2e- \rightarrow Pb(s)$

Write the overall oxidation-reduction reaction. Which metal is oxidized and which is reduced?

32. Calculate the enthalpy change (ΔH) for the *overall* oxidation-reduction reaction written above. What is ΔG for this reaction? Is this reaction spontaneous?

We know from earlier study of thermodynamics that a reaction is spontaneous if the change in free energy for the process is negative (i.e., ΔG= ΔH -T ΔS <0). For the redox reactions under consideration here (i.e., A(s) + B²⁺(aq) ⇌ A²⁺(aq) + B(s)), the entropy change is small, so the reaction is under control of the enthalpy change (ΔG ≅ ΔH).

Consider this...

By convention, the driving force for a particular electrochemical reaction is quantified by the Standard Reduction Potential (E°). Each redox pair (e.g., Zn(s) and Zn^{2+}(aq)) has its own intrinsic standard reduction potential; the more positive the potential, the greater the species' affinity for electrons and tendency to be reduced. A more positive E° means there is a greater tendency for reduction to occur. Standard reduction potentials – as shown below in Table 1 – are the potentials for the specific situation in which all species are present *in the standard state,* that is at 1 M concentration for solution species and 1 bar pressure for gases. The small ° symbol indicates the value is a standard state value. Tables listing E° values at 25°C can be found in many textbooks and reference books.

Table 1 *Standard Reduction Potentials*

Reaction	E°(Volts)
$Ag^+ + e- \rightleftharpoons Ag(s)$	0.799
$Cu^{2+} + 2e- \rightleftharpoons Cu(s)$	0.337
$2H^+ + 2e- \rightleftharpoons H_2(g)$	0.000
$Pb^{2+} + 2e- \rightleftharpoons Pb(s)$	−0.126
$Cd^{2+} + 2e- \rightleftharpoons Cd(s)$	−0.403
$Zn^{2+} + 2e- \rightleftharpoons Zn(s)$	−0.763

Key Questions

33. Which cation, Zn^{2+} or Pb^{2+}, is easier to reduce? Explain how you decided.

34. Based on your answer to Q29 where the energetics of the oxidation process were explored, which metal, Zn or Pb is easier to oxidize.

35. Are your answers to Q33 and Q34 consistent with one another?

36. Does the microscopic view of the interface (Q19) support or refute the trend in the standard reduction potentials (Q31)?

37. Is Cd^{2+} easier to reduce than Zn^{2+}? Than Pb^{2+}?

Is Cd metal easier to oxidize than Zn? Than Pb?

Consider this...

Reduction reactions, such as those listed in Table 1, do not occur independently. The electrons consumed in the reduction reaction must be generated in a coupled oxidation reaction. Oxidation-reduction reactions are based on the combination of two half-reactions (an oxidation and a reduction). Since potentials can only be measured as differences, a zero point is arbitrarily assigned to the reaction $2H^+ + 2e- \rightleftharpoons H_2(g)$, which occurs at the standard hydrogen electrode (SHE). Therefore, E° values in Table 1 are relative to the standard hydrogen electrode, which is defined as having E° equal to 0 Volts.

Combining two half-reactions can result in construction of an electrochemical cell. The difference in free energy $\Delta G°$ between the products and reactants $G°_{prod} - G°_{react}$ determines whether the reaction is favored to proceed as written.

$Zn(s) \rightleftharpoons Zn^{2+}(aq) + 2e-$
$Pb^{2+}(aq) + 2e- \rightleftharpoons Pb(s)$
$Pb^{2+}(aq) + Zn(s) \rightleftharpoons Zn^{2+}(aq) + Pb(s)$

The potential difference measured between the two electrodes, $E°_{cell}$, is also a measure of the extent to which the reaction is favored to proceed as written. The value of $E°_{cell}$ is determined using the standard reduction potentials: $E°_{cell} = E°_{red} - E°_{ox}$, where $E°_{red}$ is the standard reduction potential value (E°) for the reaction undergoing reduction and $E°_{ox}$ is the E° value for the reaction undergoing *oxidation*. Note that the calculation of $E°_{cell}$ is a *difference*, that is a subtraction.

There are other ways of calculating $E°_{cell}$ that you may have learned in earlier chemistry courses. We chose to utilize this method here as it strictly emphasizes the potential *difference* between two electrodes.

Key Questions

38. Using the values in Table 1, calculate $E°_{cell}$ for the oxidation-reduction reaction:

$Pb^{2+}(aq) + Zn(s) \rightleftharpoons Zn^{2+}(aq) + Pb(s)$

39. Based on your answer to Q32 above, is this reaction spontaneous as written?

Recall that free energy ($\Delta G°$) and $E°_{cell}$ are directly related through the equation $\Delta G° = -nFE°_{cell}$, where n is the number of electrons exchanged in the redox reaction and F is Faraday's constant (96485 C/mole e-). Thus, the cell potential $E°_{cell}$ for an oxidation-reduction reaction can be calculated and used to assess the spontaneity of the reaction.

40. Using $E°_{cell}$, calculate $\Delta G°$ for the reaction in Q38.

41. Compare your answers in Q32 and Q40.

42. Write the reverse reaction (i.e., reacting Pb(s) with Zn^{2+}) to the overall oxidation-reduction reaction in Q38. Calculate the $E°_{cell}$ for the reverse reaction. Is the reverse reaction spontaneous?

43. Imagine two situations. In one beaker, a piece of Cu metal is placed into an aqueous solution of $AgNO_3$, while in the second beaker, a piece of Ag metal is placed into an aqueous solution of $Cu(NO_3)_2$.

    a) Divide the two beakers among group members. Each subgroup should write the *overall* oxidation-reduction reaction for the reaction occurring in their assigned beaker. Come to consensus on the two reactions occurring in the two beakers.

    b) Which reaction in part a is spontaneous as written?

    c) Which of the two metal ions, Cu^{2+} or Ag^+, is more likely to be reduced? Explain your reasoning.

44. How has this activity extended your conceptual picture of electrochemical processes/reactions over what you attained in general chemistry?

45. What skill(s) did you gain in interpreting microscopic representations of electrochemical cells during this activity?

Applications

46. A drop of Hg metal is placed in a 1M $Hg_2(NO_3)_2$ solution. Describe what happens on a microscopic scale to the Hg metal and the Hg_2^{2+} ions. Draw a microscopic picture showing what happens and sketch a graph of charge vs. distance for this metal-solution interface. *Note: at this interface, reduction is the favored process.*

47. Considering Figures 2 and 3, in your own words describe the microscopic-level processes in terms of equilibrium that occurs in a beaker containing a metal electrode and its corresponding metal ion in solution.

48. Write the electrochemical equilibrium reaction when a strip of Cd metal is placed into a 1M $Cd(NO_3)_2$ solution. Is oxidation or reduction the favored process? What is the charge at the surface of the Cd electrode? What is the charge in the solution in the vicinity of the Cd electrode? Sketch a graph of charge vs. distance for this metal-solution interface. *Hint: Where does the Cd half-reaction fall in Table 1?*

49. If the standard reduction potential ($E°$) is +1.507V for the following reaction, is reduction or oxidation favored?

$$MnO_4^- + 8H^+ + 5e- \rightleftharpoons Mn^{2+} + 4H_2O$$

50. The charge at the surface of a metal electrode immersed in a salt solution of the corresponding metal nitrate as a function of distance from the electrode surface is shown below. Draw a microscopic-level picture of the electrode surface and the solution in the vicinity near the electrode. Is the metal or metal ion side of the equilibrium favored?

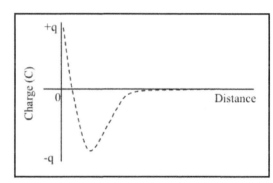

51. Four metals, A, B, C, and D, exhibit the following properties:

a) Only A and C react with 1.0 M HCl to give $H_2(g)$.

b) When C is added to solutions of the ions of the other metals, metallic B, D, and A are formed.

c) Metal D reduces B^{n+} to give metallic B and D^{n+}.

Based on the information above, arrange the four metals in order of **increasing** tendency toward oxidation.

52. The objective of this problem is to identify an unknown metal. The metal may be Fe, Mg, Ni, Pb, Sn or Zn. A microscopic-level picture of the surface of this metal when placed in a solution containing the metal nitrate is shown at the right. Water molecules are omitted for clarity. This metal is also placed into solutions of all the metal nitrates corresponding to the possible metals. Reaction occurs when the metal is placed in $Pb(NO_3)_2$ and $Sn(NO_3)_2$ solution only. Identify the mystery metal.

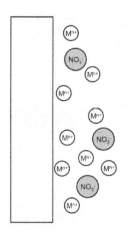

53. Thermodynamic data for the formation of $Cu^{2+}(aq)$ from Cu(s) appears in the table below. From this information, draw an energy diagram showing the three steps associated with the values in the table. What is the enthalpy of reaction for this overall reaction? Comparing the results for Cu with those for Zn and Pb (Q28), is oxidation favored in Cu to a greater or lesser extent than in Zn and Pb?

Reaction	ΔH_{sub} (kJ)	IE (1+ or 2+) (kJ)	ΔH_{hyd} (kJ)	ΔH_{rxn} (kJ)
$Cu^{2+} + 2e^- \rightleftharpoons Cu(s)$	338	2703	-2061	

54. The electrochemical cell shown here has a zinc electrode and a copper electrode. The aqueous solution into which both electrodes are placed originally contains 1 M $Zn(NO_3)_2$ and 1 M $Cu(NO_3)_2$. Sketch the interface between each electrode and the solution similar to Figure 2. Include both a "before" and "after" picture for the Zn and the Cu electrode. What reactions will occur at each metal surface?

55. A piece of silver metal is placed into a solution containing $Fe(NO_3)_3$. Does a spontaneous reaction occur? Sketch the interface between the silver metal surface and the solution.

56. Electromotive force (emf) is the work done by a source on an electrical charge. Charges are moved from lower electrical potential to higher electrical potential. A capacitor can be used to demonstrate emf because a capacitor is designed to store opposite charges on two parallel plates. After a battery is used to build up charge on the two plates (charge the capacitor), electrons can be made to flow from the negatively charged plate to the positively charged plate by connecting the two plates with a wire.

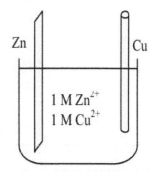

Electrochemical cell with a zinc electrode and a copper electrode placed into a solution of $Zn(NO_3)_2$ and $Cu(NO_3)_2$.

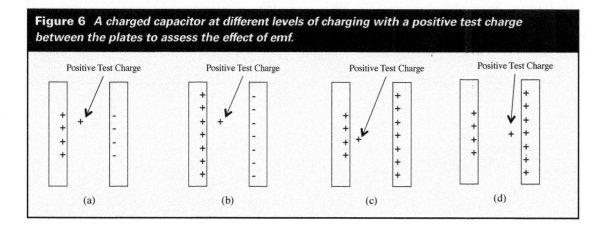

Figure 6 *A charged capacitor at different levels of charging with a positive test charge between the plates to assess the effect of emf.*

Figure 6(a) and 6(b) demonstrate the emf on a positive test charge residing between the two plates of a capacitor. In figure 6(a), there is less charge build-up on the plates than in Figure 6(b), and in Figure 6(c), the charge is the same on the two plates with different magnitudes of charge on opposite plates.

a. In Figure 6(a), which direction does the positive test charge want to travel?

To the positive plate or to the negative plate

b. In Figure 6(b), which direction does the positive test charge want to travel?

To the positive plate or to the negative plate

c. Compare figures 6(a) and 6(b).

   i. In which figure (a) or (b) is the electrostatic attraction (emf) on the positive test charge greater? Why?

   ii. In which figure (a) or (b) does the positive test charge require more energy to move toward the positive electrode? How does this compare to your answer in part a?

   iii. In which figure (a) or (b) would more energy be released if the positive test charge were allowed to move in the direction it moves spontaneously? Why?

d. List the similarities between figures 6(a) and 6(b) with figure 6(c) and 6(d). List the differences between the same pairs of figures.

e. In Figures 6(c) and 6(d), which direction does the positive test charge want to move?
Explain your reasoning. To the left or to the right

f. In which figure (c) or (d) would more energy be released if the positive test charge were allowed to move in the direction it moves spontaneously? Why?

g. Does the positive test charge experience a higher potential in figure 6(c) or 6(d)? Explain your reasoning.

h. Is the emf on the positive test charge in figure 6(c) larger or smaller than in figure 6(b)?

57. Suppose the charge vs. distance graph for a particular metal immersed in an electrolyte solution (ionic strength=0.050 M) has the form shown in the figure. How will the curve change if the ionic strength of the solution is dropped to 0.010 M? What if it is increased to 0.10 M? Explain the rationale behind your conclusions.

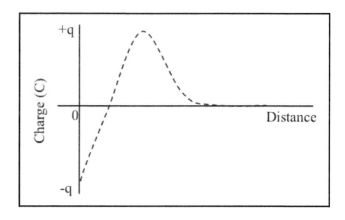

Electrochemistry: Calculating Cell Potentials

Learning Objectives

Students should be able to:

Content

- Apply the criteria for an electrochemical cell in the standard state to determine if a cell is initially in the standard state.

- Predict how the ion concentrations in an electrochemical cell will change as an electrochemical cell circuit is completed and allowed to run.

- Determine E_{cell} for a given electrochemical cell and decide whether the reaction is spontaneous in the forward or reverse direction.

Process

- Interpret drawings of electrochemical cells. (Information Processing)

- Compare and contrast different electrochemical cells. (Critical Thinking)

Prior knowledge

- Definitions of oxidation, reduction, anode, cathode, voltmeter, etc.

- Identifying oxidation and reduction processes.

- Identifying chemical reactions occurring at each electrode in an electrochemical cell.

- Calculating the cell potential for a simple electrochemical cell under standard state conditions.

- Determining if an oxidation-reduction reaction is spontaneous.

Further Reading

- Harris, D.C. 2010. *Quantitative Chemical Analysis,* 8th Edition, W.H. Freeman: USA, Sections 14-1 through 14-3, pp.309-14.

- Skoog, D.A., D.M. West, F.J. Holler, S.R. Crouch, 2004. Fundamentals of Analytical Chemistry, 8th Edition, Thompson Brooks/Cole: USA, Sections 18B, 18C-1 through 18C-3, p.496-508.

PGLANA004023a

Notes

- Table 1 provided at the end of the activity is to be used for Questions 17, 18 and 25. It is helpful to keep the table separate from the activity to add information to it.

- Table 2 provided at the end of the activity is to be used for Questions 32, 33, and 37.

- A table of standard reduction potentials (E°) is provided at the end of the activity in Q50 for when E° is needed.

Author

Christine Dalton and Mary Walczak

Section 1: Determining if an electrochemical cell is in the standard state

Consider this...

In thermodynamics, the *Standard State* of a material is a reference point used to calculate the material's properties. A gas in the standard state has a pressure of 1 bar and is at 25°C, and the standard state of solids and liquids is the pure form at 1 bar and 25°C. Solutions are in the standard state when they have an activity of 1 M at 25°C. Remember $A = [M]\gamma$, where A is the activity and γ is the activity coefficient. In analytical electrochemistry, concentrations of analyte are often in the millimolar range, but supporting electrolyte concentrations may be high enough for activity coefficients to deviate from 1. For purposes of this activity, we will make the approximation that activities are equal to concentrations.

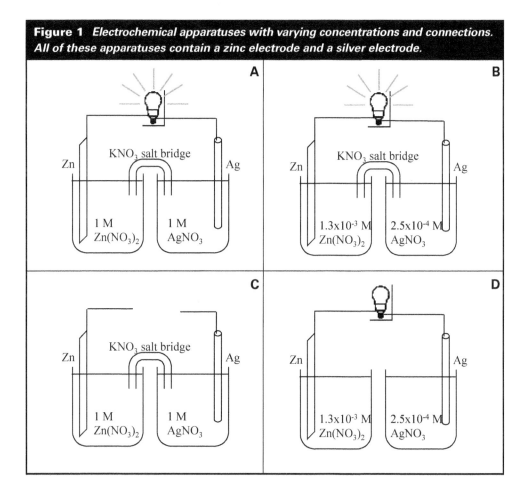

Figure 1 *Electrochemical apparatuses with varying concentrations and connections. All of these apparatuses contain a zinc electrode and a silver electrode.*

Key Questions

1. What conditions must exist for an electrochemical apparatus to be considered in the "standard state?"

2. Circle the electrochemical apparatuses in Figure 1 that meet the conditions of the standard state.

3. Figure 1 illustrates four electrochemical apparatuses. List the components that appear in these electrochemical apparatuses.

4. An electrochemical cell is distinguished from an "apparatus" in that a cell requires a complete electrical circuit. Place a star next to the apparatuses in Figure 1 that have complete electrical circuits and are therefore electrochemical cells. Compare your annotated Figure 1 with another group and resolve any differences.

5. The electrochemical "apparatuses" NOT starred in Q4 are not electrochemical cells. Explain for each why they are not cells.

6. Consider electrochemical cell A in Figure 1 (reproduced here). Given that the forward reaction in Figure 1A is spontaneous for this electrochemical cell:

- Write the two half-cells occurring in this cell.

- Label the anode (where oxidation takes place) and cathode (where reduction takes place).

- Show the direction of spontaneous current flow in the cell.

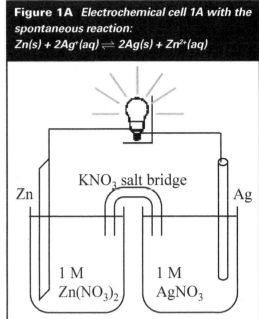

Figure 1A *Electrochemical cell 1A with the spontaneous reaction:*
$$Zn(s) + 2Ag^+(aq) \rightleftharpoons 2Ag(s) + Zn^{2+}(aq)$$

Section 2: Ion concentrations in electrochemical cells

Consider this...

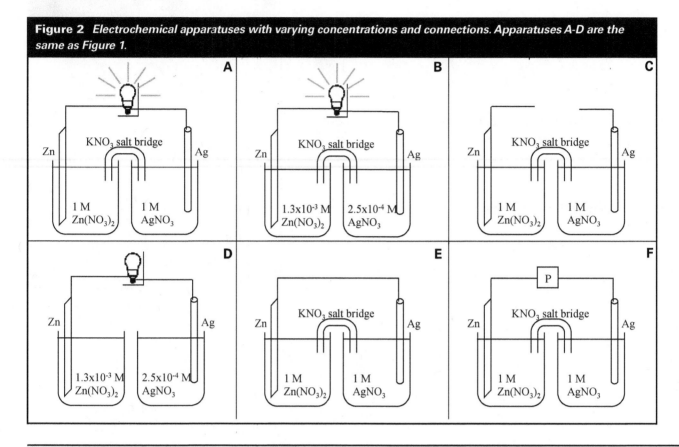

Figure 2 *Electrochemical apparatuses with varying concentrations and connections. Apparatuses A-D are the same as Figure 1.*

Key Questions

7. Consider electrochemical cell A in Figure 2. Note that the light bulb in the external circuit is illuminated. List any observations that would lead you to conclude that an electrochemical reaction occurs in this cell.

8. Consider electrochemical cell E in Figure 2. What observations lead you to conclude whether or not an electrochemical reaction occurs in this cell? Check your answer with another group.

9. Consider cells A and E only. Both of these cells are shown under conditions where the forward reaction (written in Q6) is spontaneous. At what point does the spontaneous electrochemical reaction occurring in cells A and E stop?

Apparatus F in Figure 2 contains a potentiometer, labeled "P." Today, multimeters, an example of which is shown at the right, have largely replaced potentiometers in chemistry laboratories. While multimeters can operate in several modes (e.g., measuring current, resistance, AC voltage or DC voltage), when DC voltage is selected it functions as a potentiometer—a high resistance voltmeter—and measures the potential (voltage) difference between two points. In the case of these cells it is the potential difference between the two electrodes.

http://upload.wikimedia.org/wikipedia/commons/thumb/a/a6/Digital_Multimeter_Aka.jpg/ 220px-Digital_Multimeter_Aka.jpg

10. What does "high resistance" mean? Resistance to what?

11. Recall the requirements for an apparatus to be an electrochemical cell. Place a star next to the apparatuses in Figure 2 that meet the requirements and are therefore electrochemical cells. Compare your annotated Figure 2 with another group and resolve any differences.

12. Consider electrochemical apparatus F in Figure 2. Does current flow in this electrochemical cell?

13. In electrochemical apparatus F, the electrochemical reaction does not proceed. Which of the following explains why the reaction does not proceed? Justify your answer.

(a) The electrochemical circuit is not complete.

(b) The potentiometer does not allow current to flow.

(c) The forward reaction is not spontaneous.

14. Consider the electrochemical cells at the right with a potentiometer and a light bulb in series. The light bulb does not illuminate in the cell on the left. What is the most likely explanation for this behavior?

(a) The circuit is incomplete.

(b) The resistance is too high.

(c) The bulb is burned out.

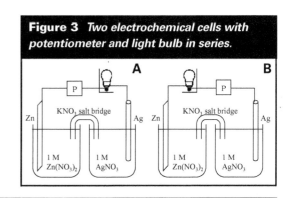

Figure 3 *Two electrochemical cells with potentiometer and light bulb in series.*

15. The light bulb does not illuminate in the other electrochemical cell either. Is this observation consistent with your explanation in the previous question? Explain your reasoning.

16. A potentiometer is used to read the voltage difference between two half-cells. What does "potential" refer to with a potentiometer (potential meter)?

17. The electrochemistry apparatuses A-F are grouped together in Table 1 at the end of the activity.

- Verify your answers to Q2 and Q4 by comparing to Table 1 entries for apparatuses A-D.

- Identify whether E and F are in the standard state and add your answers to Table 1.

- Using Q11, determine if E and F have complete circuits and add your answers to Table 1.

18. Consider again the collection of electrochemical apparatuses in Figure 2. For which of these cells do electrons flow freely (appreciable current) from the anode to cathode? Add your answers to the third row of Table 1.

Questions 19 through 23 refer to electrochemical cell A in Figure 2. Recall that the spontaneous reaction for this cell is the forward reaction: $Zn(s) + 2Ag^+(aq) \rightleftharpoons 2Ag(s) + Zn^{2+}(aq)$

---

19. What happens to the zinc metal after the spontaneous forward reaction has proceeded to a significant extent? Defend your choice.

(a) The Zn electrode increases in mass.

(b) The Zn electrode maintains its original mass.

(c) The Zn electrode has an overall smaller mass.

20. Consider the silver half-cell. Which of the following statements is true after the spontaneous forward reaction has proceeded to a significant extent?

(a) The Ag electrode decreases in mass.

(b) The $[Ag^+]$ decreases.

(c) The Ag electrode is unchanged as this reaction proceeds.

---

21. Based on your answers to Q19 and Q20, how does the concentration of cations (i.e., Ag^+ and Zn^{2+}) in solution change after the spontaneous forward reaction has proceeded to a significant extent?

The number of Ag^+ in the right half-cell:　　　　increases　　　　stays the same　　　　decreases

The number of Zn^{2+} in the left half-cell:　　　　increases　　　　stays the same　　　　decreases

22. Consider the graph below that shows the concentration of Zn^{2+} in cell A as a function of time.

(a) Why does one of the lines go up, while the other line goes down?

(b) Are the rates of change for the two lines different? Provide justification for you answer.

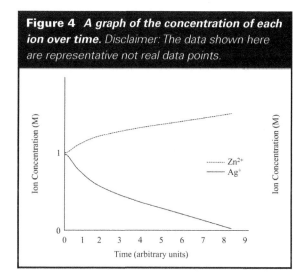

Figure 4 *A graph of the concentration of each ion over time.* Disclaimer: The data shown here are representative not real data points.

23. Recall the conditions for standard state defined on page 1. At which point(s) in time is electrochemical cell A in the standard state?

24. Electrochemical Cell F from Figure 2 appears below. Sketch the concentration profile for this cell.

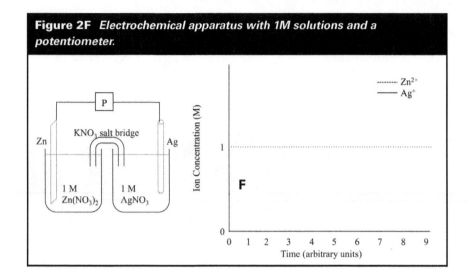

Figure 2F *Electrochemical apparatus with 1M solutions and a potentiometer.*

25. Consider all the electrochemical apparatuses in Figure 2. Which of these have concentrations that DO NOT change over time. Add your entries to the fourth row of Table 1.

26. How do you know if an electrochemical cell starts at 1 M and remains at 1 M?

Section 3: Calculating cell potentials

Consider this...

The potential difference between two electrodes in an electrochemical cell, E_{cell}, is determined by the Nernst equation:

$$E_{cell} = E^{\circ}_{cell} - \frac{RT}{nF}\, lnQ = E^{\circ}cell - \frac{0.05916V}{n}\, logQ$$

where E°_{cell} is the cell potential measured under standard conditions, n is the number of moles of electrons transferred in the electrochemical reaction, and Q is the reaction quotient for the forward reaction as written. Recall that the Nernst equation is frequently expressed in terms of base 10 logarithms for a cell at 25°C, in which case the ln term becomes the log term.

Key Questions

27. Write the Nernst equation expression for electrochemical cell F from Figure 2 (and shown here), given that the forward reaction below is spontaneous for this electrochemical cell.

$$Zn(s) + 2Ag^{+}(aq) \rightleftharpoons 2Ag(s) + Zn^{2+}(aq)$$

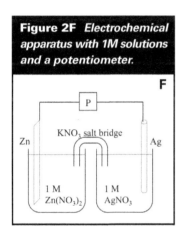

Figure 2F *Electrochemical apparatus with 1M solutions and a potentiometer.*

28. Look up the standard reduction potentials for the two half-reactions in electrochemical cell F in the table of standard reduction potentials provided in Q50 in the Applications section of this activity. Calculate E°_{cell} for the electrochemical reaction in Q27 under the conditions of cell F?

29. Calculate E_{cell} for electrochemical cell F using the Nernst equation expression written in Q27 and the E°_{cell} value from Q28.

30. The Nernst equation cannot be used to calculate Ecell for the other five electrochemical apparatuses in Figure 2. For each electrochemical apparatus A-E specify the reason that Nernst equation can't be used.

A:

B:

C:

D:

E:

31. Are there any points in time that the Nernst equation can be used for any of the electrochemical apparatuses A-E? If so, indicate when it can be used.

Consider this...

The following electrochemical cells contain a Ag electrode immersed in $AgNO_3$ solution and a Pt electrode in a solution containing both Fe^{2+} and Fe^{3+}. We will consider each of these four cells separately.

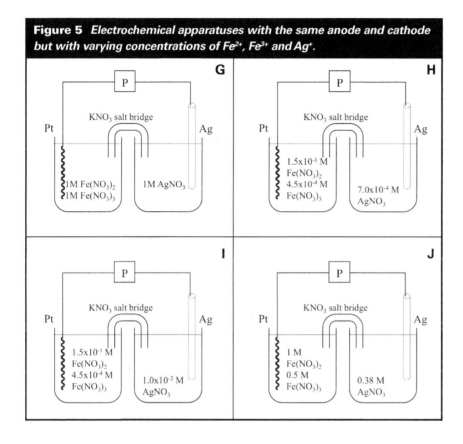

Figure 5 *Electrochemical apparatuses with the same anode and cathode but with varying concentrations of Fe^{2+}, Fe^{3+} and Ag^+.*

32. The reaction for electrochemical cell G is $Ag^+ + Fe^{2+} \rightleftharpoons Ag(s) + Fe^{3+}$, which is spontaneous in the forward direction. Does E^0_{cell} indicate if this reaction at standard state is spontaneous? Use the Nernst equation to determine E_{cell} and for cell G. Add these values to Table 2 at the end of the activity.

33. Divide the remaining three cells H-J among group members. Use the Nernst equation to determine E_{cell} and E°_{cell} for cells H-J. Add these values to Table 2.

34. Look at the E_{cell} and $E°_{cell}$ values in Table 2 for cells G-J.

   a) Are all $E°_{cell}$ values equal? Why?

   b) Is E_{cell} always greater than or equal to $E°_{cell}$?

35. Using the Nernst equation, specify when

   a) $E_{cell} = E°_{cell}$

   b) $E_{cell} < 0$

   c) $E_{cell} > 0$

36. Using your prior knowledge of thermodynamics, what happens to a reaction when $E_{cell} < 0$ (i.e., $\Delta G > 0$)? Use "spontaneous" in your answer.

37. For cells H-J in Figure 5 decide which of the three statements about the reaction applies to each case. Mark your answers in Table 2.

   A. The reaction proceeds in the forward direction: $Ag^+ + Fe^{2+} \rightleftharpoons Ag(s) + Fe^{3+}$

   B. The reaction proceeds in the reverse direction: $Ag(s) + Fe^{3+} \rightleftharpoons Ag^+ + Fe^{2+}$

   C. The reaction proceeds in neither direction because the reaction is at equilibrium

38. Support or refute this statement:

E°_{cell} *can be always be used to indicate the spontaneity of an electrochemical cell.*

Self-Assessment Questions

39. List one misconception that you had about electrochemical cells before doing this activity that has now been addressed.

40. How do you judge whether an electrochemical cell is in the standard state?

Applications

Note: In some instances, you will have to look up the standard reduction potential in the textbook to be able to complete the problem.

41. The electrochemical cells shown here have a zinc anode and a platinum electrode instead of silver in the other cells in this activity. The Zn electrode is immersed in a solution containing 1 M $Zn(NO_3)_2$, while the cathode is composed of a platinum electrode immersed in 1 M HNO_3 with H_2 (g) bubbled into the solution through a tube. The two metal electrodes are connected by a wire with a potentiometer (K) or light bulb (L) in the line. The KNO_3 salt bridge completes the electrical circuit between the separated half-cells.

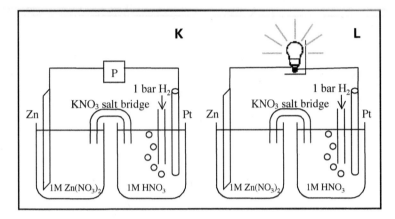

(a) Draw the path of H^+, Zn^{2+} and e- movement/flow on each of the figures above, as appropriate.

(b) Is Apparatus K in the standard state at any point after the circuit is complete? Is Apparatus L? Specify the time(s) at which each apparatus is the standard state.

(c) Calculate E_{cell} for apparatus K.

(d) Why can't E_{cell} be calculated for apparatus L?

42. Given this reaction:

$$Zn(s) + Cu^{2+}(aq) \rightleftharpoons Zn^{2+}(aq) + Cu(s)$$

(a) Sketch an electrochemical cell (with complete circuit) in the standard state.

(b) Calculate E_{cell} for this cell.

(c) As you have drawn the cell, will it remain in standard state after it is initially connected?

43. Calculate E_{cell} for this reaction if the solution concentrations are

$[Cr^{2+}] = 2.4 \times 10^{-2}$ M and $[Fe^{2+}] = 1.8 \times 10^{-3}$ M.

$$Cr(s) + Fe^{2+}(aq) \rightleftharpoons Cr^{2+}(aq) + Fe(s)$$

44. Consider electrochemical apparatus M at the right.

    (a) What is the overall spontaneous reaction for electrochemical apparatus M? Calculate E_{cell}.

    (b) How would E_{cell} change if the potentiometer were removed and replaced with a wire connecting the Pb and Fe electrodes?

    (c) Suppose the cell with the wire connection is allowed to run until the potential difference between the two electrodes was 0.28 V. At this point in the discharge of the electrochemical cell what are $[Fe^{2+}]$ and $[Pb^{2+}]$?

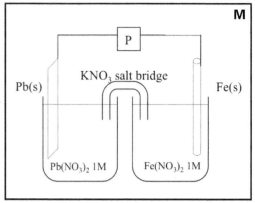

45. Compare and contrast an electrochemical cell that has a light bulb connecting the two electrodes (like Electrochemical Cell A in Figure 2) and one with a potentiometer connecting the two electrodes (like Electrochemical Cell F in Figure 2).

46. The graph shows a concentration vs. time profile for three different half-cells with the reaction $M^{n+} + n\,e^- \rightleftharpoons M(s)$. The concentration refers to the M^{n+} ion in solution as the cell is set up as part of an electrochemical cell and allowed to run.

    (a) For each cell 1-3, specify whether the cell is ever in the standard state. If it is, then specify at which times during the experiment it is in the standard state.

    (b) Imagine the half-cells 1-3 are connected to another half-cell with a completed circuit. For which of the half-cells 1-3 does current flow? Explain how you decided.

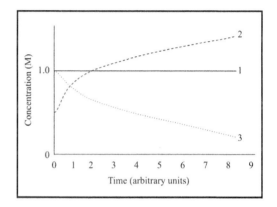

    (c) Since the half-reaction corresponding to each of these concentration profiles is $M^{n+} + n\,e^- \rightleftharpoons M(s)$, in which cases 1-3 does the reaction occur as written (i.e., reduction of the metal ion to solid metal)?

47. You decide to construct an electrochemical cell to power your fan. Your fan requires a voltage of 1.50 V in order to operate. You have the following three half-cells to combine to form a complete cell generating a voltage sufficient to power your fan. In each half-cell the metal ion concentration is 1.0 M.

Reaction	E° (Volts)
$Ag^+ + e^- \rightleftharpoons Ag(s)$	0.799
$Pb^{2+} + 2e^- \rightleftharpoons Pb(s)$	-0.126
$Zn^{2+} + 2e^- \rightleftharpoons Zn(s)$	-0.763

(a) Which pair of half-cells do you choose?

(b) If the clock draws 1.0 mA of current, how long will it run until the voltage drops below 1.50 V?

(c) Does it matter which electrode is connected to the positive terminal of your fan? Why or why not?

48. Suppose the wire connecting the Zn and Ag electrodes attached to the light bulb is removed from Electrochemical Cell A and the leads are reversed so the end originally attached to the Zn electrode is attached to the Ag electrode and vice versa. Will the light bulb produce light? Explain your reasoning.

49. If the potentiometer (which measures voltage) in Electrochemical Cell F is connected in the opposite direction (i.e., the leads are switched) how will the reading on the meter change from the value it had originally?

50. The Table below lists Standard Reduction Potentials for several half-reactions. For what concentrations of solution species are these $E°$ values valid?

Standard Reduction Potentials for Selected Half-Reactions	
Reaction	**E° (Volts)**
$MnO_4^- + 8 H^+ + 5 e- \rightleftharpoons Mn^{2+} + 4 H_2O$	1.507
$Ag^+ + e- \rightleftharpoons Ag(s)$	0.799
$Fe^{3+} + e- \rightleftharpoons Fe^{2+}$	0.771
$Cu^{2+} + 2e- \rightleftharpoons Cu(s)$	0.339
$2H^+ + 2e- \rightleftharpoons H2(g)$	0.000
$Pb^{2+} + 2e- \rightleftharpoons Pb(s)$	−0.126
$Cd^{2+} + 2e- \rightleftharpoons Cd(s)$	−0.402
$Fe^{2+} + 2e- \rightleftharpoons Fe(s)$	−0.44
$Zn^{2+} + 2e- \rightleftharpoons Zn(s)$	−0.762
$Cr^{2+} + 2e- \rightleftharpoons Cr(s)$	−0.89

51. Using the axes below, sketch how the potential difference between the Cu and Zn electrodes will change as a function of time for Electrochemical Cell N and for Electrochemical Cell O, also shown below.

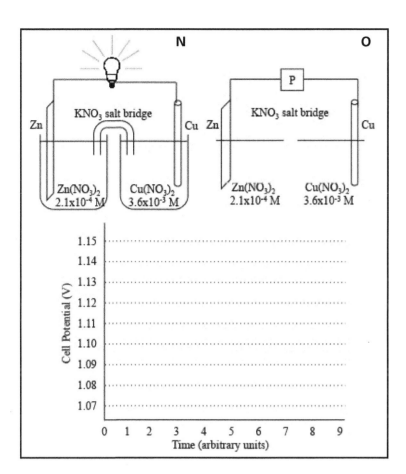

Table 1

	A	B	C
Electrochemical Apparatus	Zn, KNO₃ salt bridge, Ag, 1 M Zn(NO₃)₂, 1 M AgNO₃	Zn, KNO₃ salt bridge, Ag, 1.3x10⁻³ M Zn(NO₃)₂, 2.5x10⁻⁴ M AgNO₃	Zn, KNO₃ salt bridge, Ag, 1 M Zn(NO₃)₂, 1 M AgNO₃
Standard State Concentrations?	Yes	No	Yes
Complete Circuit?	Yes	Yes	No
Electrical Current?			
Concentrations Fixed?			

	D	E	F
Electrochemical Apparatus	Zn, Ag, 1.3x10⁻³ M Zn(NO₃)₂, 2.5x10⁻⁴ M AgNO₃	Zn, KNO₃ salt bridge, Ag, 1 M Zn(NO₃)₂, 1 M AgNO₃	Zn, P, KNO₃ salt bridge, Ag, 1 M Zn(NO₃)₂, 1 M AgNO₃
Standard State Concentrations?	No		
Complete Circuit?	No		
Electrical Current?			
Concentrations Fixed?			

Table 2

		G	H
Electrochemical Apparatus			
E°_{cell} (V)	E_{cell} (V)				
The reaction proceeds in the forward direction: $Ag^+ + Fe^{2+} \rightleftharpoons Ag(s) + Fe^{3+}$			
The reaction proceeds in the reverse direction: $Ag(s) + Fe^{3+} \rightleftharpoons Ag^+ + Fe^{2+}$			
The reaction proceeds in neither direction because the reaction is at equilibrium			

		I	J
Electrochemical Apparatus			
E°_{cell} (V)	E_{cell} (V)				
The reaction proceeds in the forward direction: $Ag^+ + Fe^{2+} \rightleftharpoons Ag(s) + Fe^{3+}$			
The reaction proceeds in the reverse direction: $Ag(s) + Fe^{3+} \rightleftharpoons Ag^+ + Fe^{2+}$			
The reaction proceeds in neither direction because the reaction is at equilibrium			

The Beer-Lambert Law

Learning Objectives

Students should be able to:

Content

- Identify each term in the Beer-Lambert law and explain its effect on absorbance.

- Explain the relationship between transmittance and absorbance.

- Use Beer's Law in quantitative measurements.

Process

- Prepare and interpret graphs. (Information processing)

- Develop mathematical expressions to describe data. (Problem solving).

Prior knowledge

- Calibration curves and linear equations

Further Reading

- Harris, D.C. 2007. *Quantitative Chemical Analysis*, 7th Edition, Section 18-2, p. 380. New York: WH Freeman.

- Boyer, R.F. 2012 *Biochemistry Laboratory: Modern Theory and Techniques*, 2nd Edition, Section 3-B, pp. 67-71 and Chapter 7A, pp. 202-220. Boston: Prentice Hall.

Authors

Caryl Fish, David Langhus, and Kathleen Cornely

PGLANA00XXXXa

Consider this...

Suppose that we have a solution that absorbs light.

Suppose further, that we propose to analyze the solution using an instrument organized as shown below:

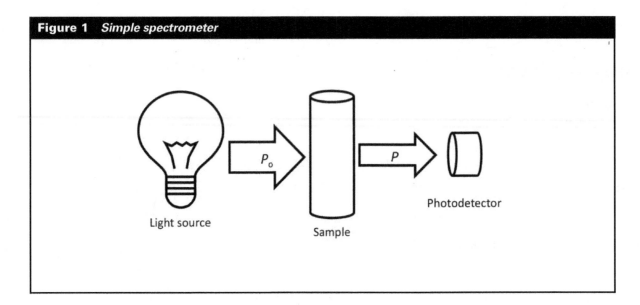

Figure 1 *Simple spectrometer*

The light source might be an ordinary light bulb, a flashlight, or the sun. The photodetector could be a simple solar cell available from Radio Shack. The light shining on the sample has power P_0 and the light that manages to get through the sample has power P. P_0 and P may be expressed in Lamberts or lumens or ergs/sec/cm^2 or any other units, as long as the same units are used for both.

Transmittance is defined as $T = P/P_0$ (or sometimes denoted as I/I_0, where I is the intensity of the light shining on the sample and I_0 is the intensity of the transmitted light).

Percent transmittance, or % T, is simply 100 T.

Key Questions

1. If some of the light is absorbed by the sample....

   Is P greater than, less than, or equal to P_0? Justify your answer.

   Is T greater than, less than, or equal to 1? Justify your answer.

2. Absorbance indicates how much light is absorbed by a sample. If sample A (below) absorbs more light than sample B, which would have the higher transmittance? Which would have the higher absorbance? Explain your reasoning.

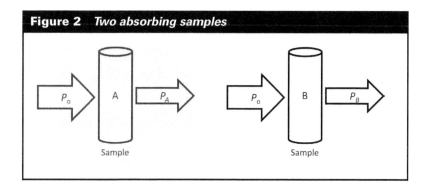

Figure 2 *Two absorbing samples*

3. Discuss with your group members what happens to transmittance when the absorbance changes. Describe the relationship between transmittance and absorbance in 2-3 complete sentences.

Consider this…

Suppose that we have six identical samples of a light-absorbing solution (in containers that do not absorb light) arranged in a line with a light source with power P_0 aimed at the first sample as shown in Figure 3:

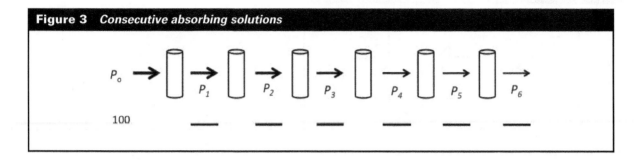

Figure 3   Consecutive absorbing solutions

Key Questions

4. If 50% of the light is transmitted through each sample and the initial power, P_0, is 100 arbitrary units, calculate values for P_1 through P_6. Fill in the blanks in the figure with these values and compare your values with your group members.

5. Using the equation for transmittance, $T = P/P_0$, each group member should calculate the total transmittance value for two different n values where n is the number of solutions through which light passes. Share your answers and plot the T values vs. n on the plot shown below in Figure 4.

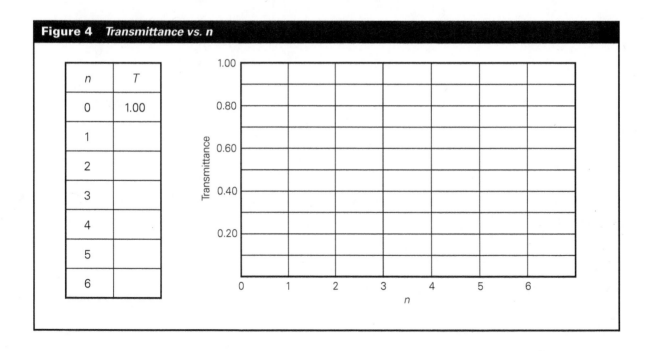

Figure 4   Transmittance vs. n

n	T
0	1.00
1	
2	
3	
4	
5	
6	

6. Does the transmittance increase or decrease as n increases? Is there a linear relationship between the transmittance and the number of samples through which the light passes?

7. Repeat the exercise above but this time plot the $-\log_{10} T$ vs. n (where n = the number of solutions through which light passes).

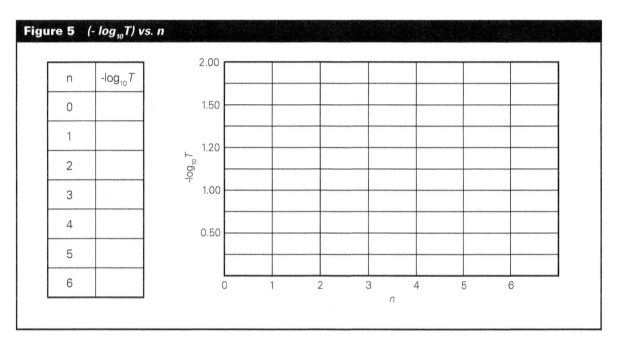

Figure 5 $(-\log_{10}T)$ vs. n

n	$-\log_{10} T$
0	
1	
2	
3	
4	
5	
6	

Does the $-\log_{10}T$ vary linearly with n?

8. Using the standard form for a linear equation ($y = mx + b$, where m is the slope and b is the intercept), write an equation for the relationship shown on the graph in Figure 5. What value should be expected for the intercept?

Now it is convenient to describe $-\log_{10}T$ as the absorbance (A) such that $A = -\log_{10}T$. Rewrite your equation in Q8 in terms of absorbance.

9. Suppose that each individual sample in the line of samples in Figure 3 is a slice of a mega-sample. In other words, instead of six successive solutions there is one large container with six slices of solution put together.

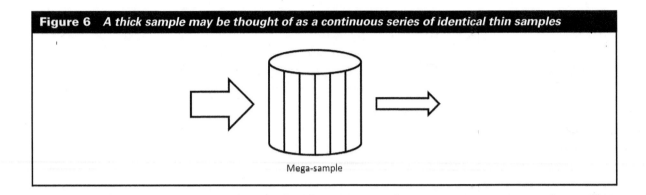

Figure 6 *A thick sample may be thought of as a continuous series of identical thin samples*

Mega-sample

Look back at your data in Figure 5, but now think of n as the number of slices in the sample. Describe the relationship between the distance the light travels through the sample and $-\log_{10} T$ (or A) in one complete sentence.

10. Using ℓ (instead of n) to represent the total distance the light travels—the path length—modify your mathematical expression in Q8 to express the relationship between path length, ℓ, and absorbance.

Consider this...

Three samples with different amounts of absorbing species (represented by ✺) are shown in Figure 7 below. The distance the light travels through each sample, the path length, ℓ, and the power of the incoming light (P_0) are the same for all three samples.

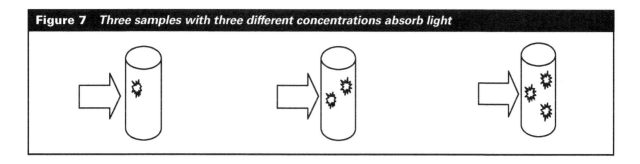

Figure 7   Three samples with three different concentrations absorb light

11. Look back at Figure 5, except now assume that n is the number of absorbing species in a given volume of solution, *i.e.*, the concentration, which we will denote as c. Discuss with your group the relationship between concentration *(c)* and absorbance. Draw arrows to represent the transmitted power *(P)* for each sample in Figure 7 above; then modify your mathematical expression in Q10 to include concentration.

The Beer-Lambert law (or simply Beer's law) describes the relationship between the absorbance, A, of a solution, the distance through which the light travels through the solution, and the concentration of the solution:

$$A = Ec\ell$$

E = absorptivity, a proportionality constant that defines the efficiency or the extent of the absorption.
c = concentration of absorbing species in the sample.
ℓ = path length of light through the sample, or distance the light travels through the sample.

---

12. Refer to the equation you wrote in Q11 and the equation for Beer's law shown in the box above. Compare the two equations and comment on the slope, *m*, and absorptivity, *E*.

Consider this...

A spectrophotometer is an instrument that measures the absorbance of a sample at different wavelengths of light. A graph of the absorbance of a sample as a function of wavelength obtained from such an instrument is called an **absorbance spectrum** and is shown in Figure 8. A 65 µM solution of bovine serum albumin (BSA) placed in a 1.00 cm diameter cuvette yields the spectrum shown in Figure 8.

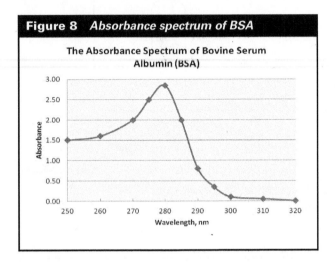

Figure 8   Absorbance spectrum of BSA

Key Questions

13. What terms in Beer's law were held constant in order to obtain the absorbance spectrum shown in Figure 8?

14. Use the absorbance spectrum in Figure 8 to answer the following questions: What is the absorbance of bovine serum albumin under these conditions at 250 nm? At 280 nm?

15. The Beer's law equation can be used to determine the concentration of a solution if E is known, by measuring the absorbance. Refer to Figure 8 and choose the best wavelength to use to determine the concentration of a BSA solution. Come to a consensus and explain your choice in terms of the sensitivity of the measurement (i.e., how well you can distinguish between solutions of low and high concentrations).

16. Discuss with your group how the data displayed in the absorbance spectrum in Figure 8 illustrate Beer's law. What term in Beer's Law must be wavelength-dependent?

Consider this...

In the biochemical literature, the absorptivity, E, for proteins is usually reported as a percent extinction coefficient (ε percent). The symbol ε is used instead of E, but ε is really another form of E. The percent extinction coefficient is defined as the proportionality constant that relates the absorbance, A, to the concentration of a 1% solution (1 g/100 mL or 10 mg/mL) in a cell with a path length, ℓ, of 1.00 cm at a specific wavelength. Sometimes the concentration is expressed as a 0.1% solution (0.1 g/100 mL or 1 mg/mL) instead of a 1.0% solution. The percent extinction coefficient is denoted as $A^{\%}_{\lambda}$ in which the percentage, either 1% or 0.1%, is shown in the superscript to the right of the A and the wavelength, λ, is shown in the subscript. The various ways of denoting the percent extinction coefficients are shown below for a typical antibody solution:

$$A^{1\%}_{280nm} = 14 \text{ or } A^{0.1\%}_{280nm} = A^{1mg/mL}_{280nm} = 1.4$$

The above notation indicates that the absorbance of a 1% solution of an antibody, measured in a 1.00 cm diameter cuvette at 280 nm, has an absorbance of 14. The absorbance of a 0.1% solution (1 mg/mL) of antibody measured under the same conditions will be 1.4.

17. Use Beer's law, $A = \varepsilon_{percent} \, c\ell$, determine the units of $\varepsilon_{percent}$ for a 1% protein solution. (Remember that absorbance, A, does not have units.)

18. In other cases, ε is expressed as the **molar extinction coefficient** and is denoted as ε_{molar}. The molar extinction coefficient is defined as the proportionality constant that relates the absorbance, A, to the concentration of a 1 M solution in a cell with a path length, ℓ, of 1.00 cm at a specific wavelength. Use Beer's law, $A = \varepsilon_{molar} \, c\ell$ to determine the units for ε_{molar}.

19. Most proteins absorb strongly at 280 nm because of the presence of tyrosine and tryptophan residues, whose side chains contain conjugated double bonds that absorb strongly at this wavelength. This information can be used to compare the molar extinction coefficients of proteins of different amino acid content. The molar extinction coefficients were determined both for a native ribonuclease enzyme and a mutant ribonuclease enzyme in which an aspartate residue was mutated to a tryptophan. Identify these proteins by their molar extinction coefficients by completing the table below (from Pace, C.N. *et al.* 1995. "How to Measure and Predict the Molar Absorption Coefficient of a Protein." Protein Science 4 (11): 2411-2423.)

Molar extinction coefficient, $M^{-1}cm^{-1}$	Protein
17,420	
21,700	

20. What are two things you did not know before but learned about Beer's Law during this activity?

21. In what ways did this activity enhance your ability to write mathematical expressions to describe relationships?

Applications

22. A spectrophotometer will often produce absorbance and % transmittance values. Which value would be easier for you to use to determine the concentration of a compound in solution using Beer's law? Explain.

23. The molar extinction coefficient, ε_{molar}, for IgG is 210,000 M^{-1} cm^{-1}. A solution of IgG is placed in a cell with a pathlength of 0.50 cm. The absorbance measured is 0.95. What is the concentration of the IgG solution?

24. You are working with a solution of BSA in the lab and need to know its concentration for a protein assay that you are running. You measure the absorbance at 280 nm in a 1.00 cm cuvette and obtain an absorbance of 1.523. The $\varepsilon_{percent}$ for BSA is reported as $A_{280nm}^{1\%}$ = 6.6. What is the concentration of the solution in mg/mL?

25. A 1% solution of tyrosinase has an absorbance of 24.9. What is the concentration of a solution of tyrosinase that has an absorbance of 0.0295?

26. A kinetics experiment is carried out in which the substrate L-DOPA (which is colorless in solution) is added to an enzyme that converts the substrate to the red-colored dopachrome product. The molar extinction coefficient of dopachrome is 3600 M^{-1} cm^{-1}. The reaction is run in a cuvette with a diameter of 2 mm. The rate of the formation of dopachrome product is measured to be 0.000831 absorbance units per minute. What is the rate of the reaction in units of M of product formed per minute?

27. The measurement of protein concentration in a solution becomes problematic when the solution contains a mixture of proteins rather than a single purified protein. Why is it not possible to use Beer's law to determine the concentration of a mixture of proteins?

28. A different application of Beer's law is used to determine the concentration of a mixture of proteins. The strategy involves adding a reagent that converts the protein to a colored compound. For example, if an alkaline copper (II) sulfate solution is added to a protein, a reaction with the peptide bonds occurs and produces Cu^+ as a product. The Cu^+ subsequently reacts with bicinchoninic acid (BCA) to form a purple complex that absorbs strongly at 562 nm. The intensity of the purple color is proportional to the protein concentration. A series of protein standards from 25-750 µg/ml were prepared, reacted with the alkaline copper/BCA solution, and then measured for absorbance. An unknown was similarly treated. The data are shown below. Using these data, calculate the protein concentration in the unknown solution.

Standard protein concentration, µg/mL	Absorbance @ 562 nm
0	0.078
25	0.120
125	0.247
250	0.427
500	0.702
750	0.940
unknown	0.850

29. Suppose that in measuring a sample of protein using an absorption spectrophotometer, you find that the absorbance is too high for the instrument to measure accurately. Use Beer's law to determine two options you might use to decrease the absorbance value.

30. Using Beer's law is a powerful method to use to determine the concentration of an unknown protein solution, but there are limitations. The method works well for monochromatic radiation, dilute solutions, and absorbing compounds that do not participate in concentration-equilibrium reactions. A student performs a colorimetric protein assay and obtains the data tabulated below. Construct a Beer's law plot using these data and determine the highest concentration at which Beer's Law applies.

[BSA], µg/mL	A @ 562 nm
0	0.000
25	0.125
125	0.640
250	0.854
500	1.432
750	1.954
1000	2.052
1500	2.120

Atomic and Molecular Absorption Processes

Learning Objectives

Students should be able to:

Content

- Relate the wavelength and energy of light absorbed to molecular and atomic transitions.

- Compare molecular and atomic absorption processes and explain differences in band width in the resulting spectra.

Process

- Compare and contrast spectra and absorption processes (critical thinking).

Prior knowledge

- Electromagnetic spectrum.

- Energy and Wavelength relationships.

- Molecular orbital terminology.

Further Reading

- Skoog, D.A., Holler, J.F., Crouch, S.R. *Principles of Instrumental Analysis,* Sixth Edition, Thomson Brooks/Cole, 2007, pp. 131-164 and 336-367.

- D.C. Harris, *Quantitative Chemical Analysis,* 7th Edition, 2007 W.H. Freeman: USA, Section 18-5, pp. 387-389.

- A.M. Bass, L.C. Glasgow, C. Miller, J.P. Jesson, and D.L. Filkin, "Temperature dependent absorption cross sections for formaldehyde (CH_2O): The effect of formaldehyde on stratospheric chlorine chemistry," Planet. Space Sci. 28, 1980, 675-679.

Authors

Caryl Fish, Ruth Riter

Consider this...

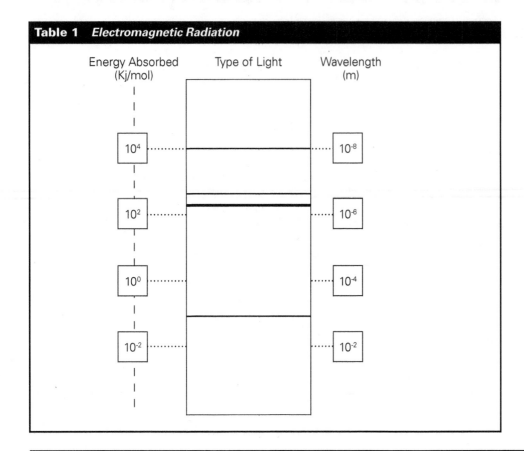

Table 1 *Electromagnetic Radiation*

Energy Absorbed (Kj/mol): 10^4, 10^2, 10^0, 10^{-2}

Type of Light

Wavelength (m): 10^{-8}, 10^{-6}, 10^{-4}, 10^{-2}

Key Questions

1. How are the energy and wavelength related in the table?

2. Recall the relationship between wavelength and frequency. Is the table consistent with E=hv? Explain your reasoning.

3. Add the types of electromagnetic radiation below to Table 1, above.

    a. Infrared light ($\lambda = 7.80 \times 10^{-7} - 1 \times 10^{-3}$ m)

    b. Ultraviolet light ($\lambda = 1 \times 10^{-8} - 3.80 \times 10^{-7}$ m)

    c. X-rays ($\lambda = 10^{-11}$ to 10^{-8} m)

    d. Microwave ($\lambda = 10^{-3}$ to 10^{-1} m)

    e. Visible light ($\lambda = 3.80 \times 10^{-7}$ to 7.80×10^{-7} m)

Consider this...

Figure 1 *Atomic Absorption of Sodium*

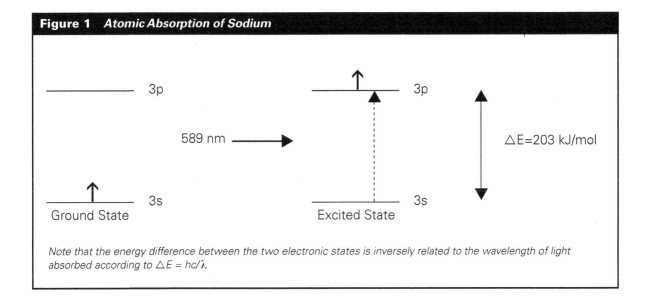

Note that the energy difference between the two electronic states is inversely related to the wavelength of light absorbed according to $\triangle E = hc/\lambda$.

Key Questions

4. In Figure 1, the outermost electron of sodium resides in which ground state atomic orbital?

5. The electron is in which atomic orbital after the atom absorbs a photon of light?

6. How much energy is absorbed in this transition?

7. Sodium absorbs light at both 589.0 and 589.6 nm due to differences in spin of the outermost electron. At this point assume equal absorbance at each wavelength. Using this information, draw a sketch that represents wavelength vs. absorbance for sodium.

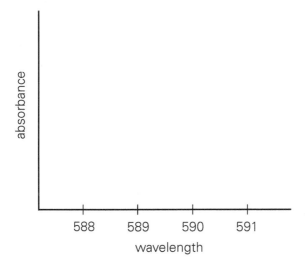

> **In atomic absorption the absorption bands are very narrow, and less than a nanometer wide.**

8. After reading the above information, compare your sketch with your group members and adjust your sketch if necessary.

Consider this...

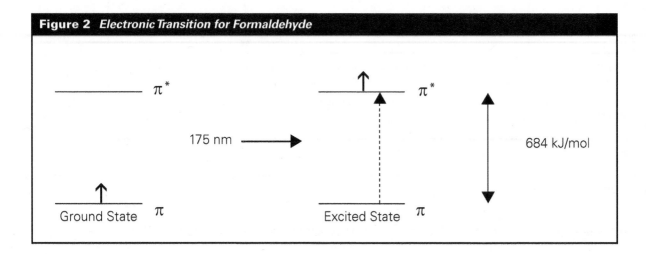

Figure 2 *Electronic Transition for Formaldehyde*

Key Questions

9. In Figure 2, what do the π and π* symbols represent?

10. In what ways is this diagram of molecular absorption different than the atomic absorption diagram?

    How is it similar?

11. Based on the wavelength, the absorption of what type of electromagnetic radiation causes the electronic transition in formaldehyde?

Consider this...

Molecules also have vibrational and rotational energy states. Below are vibrational and rotational transitions for formaldehyde.

Vibrational transition   Rotational transition

Key Questions

12. Compare the energy required for the electronic, vibrational, and rotational transitions of formaldehyde. Which requires the most energy? Which the least?

13. The energy required for the vibrational transition is about (10, 100, 1000,. or 10,000) times more than the energy required for the rotational transition.

The energy required for the electronic transition is about (10, 100, 1000 or 10,000) times more than the energy required for the vibrational transition.

14. Revisiting our original table, now add the molecular processes of electronic, vibrational, and rotational excitation to the appropriate boxes in the table below.

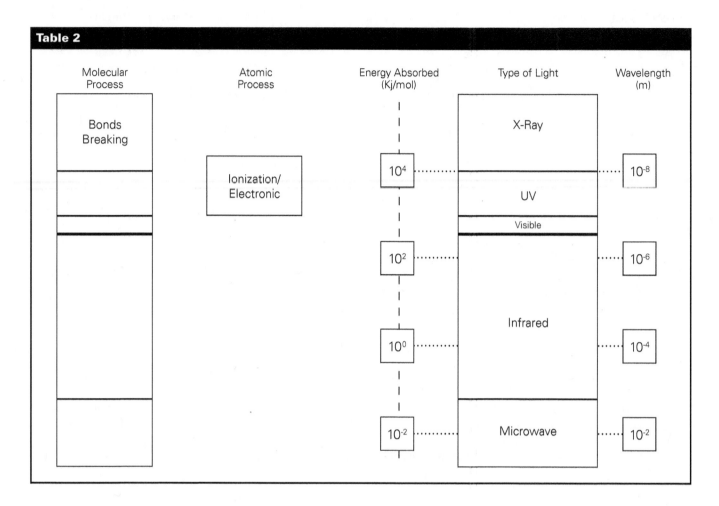

Consider this...

The different transitions in a molecule do not occur independently, but when enough energy is absorbed for an electronic transition, the vibrational and rotational transitions overlap with the electronic transitions. Since molecules can rotate and vibrate these transitions impact the spectrum. Let's assume that each electronic state of formaldehyde has 4 vibrational states. Each vibrational state would also have a number of rotational states, but the energy differences would be very small and hard to show on the diagram.

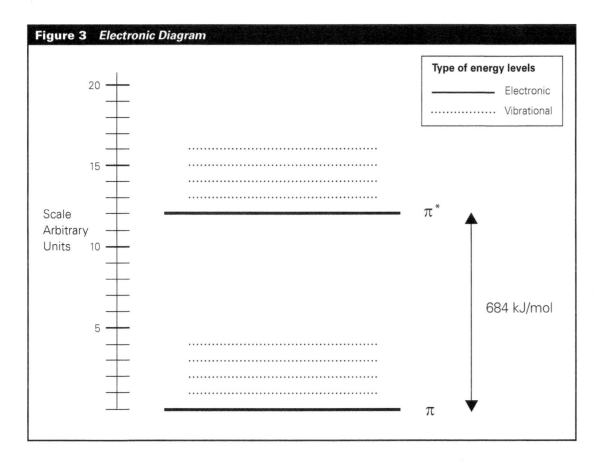

Figure 3 *Electronic Diagram*

15. Draw all the transitions from one of the π energy levels to all the π* energy levels. Assign each student in your group to start from a different vibrational level in the π electronic energy level. (For this model, do not worry about "allowed" transitions)

16. Each transition has a length. Use the scale on the side to determine the length of each of your transitions and list those below.

_____   _____   _____

_____   _____   _____

17. For your entire group, determine the number of transitions of each length and record them in the table below.

Length	Number of Transitions	Length	Number of Transitions
8		13	
9		14	
10		15	
11		16	
12		17	

18. Sketch the length vs. number of transitions on the graph below and connect the points.

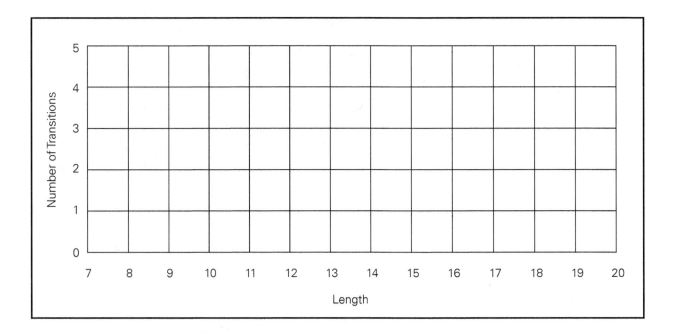

19. This sketch is analogous to one band in the UV-Vis spectrum for formaldehyde (see formaldehyde spectrum at the end of the activity if you need help). What does the length of each transition represent in the UV-Vis spectrum? What does the number of transitions represent?

> **The actual intensity at each wavelength is not due to only the number of possible transitions.**
> **Some transitions are more likely to occur and some have different intrinsic strengths.**

20. Compare the spectra drawn in Q7 and Q18. Discuss with your group the differences and similarities between a molecular spectrum and an atomic spectrum. Record a list of both differences and similarities. Be sure to consider the range of possible wavelengths and the type of molecular processes recorded.

21. Examine the gas phase spectrum of formaldehyde and gas phase transmittance spectrum of sodium at the end of this activity and add any additional similarities and differences to your lists.

22. Based on molecular processes, discuss why a spectrum for molecular absorption is much broader than a spectrum for atomic absorption. Share your explanation with another group. Record a consensus explanation in your own words.

23. What are two specific concepts that you learned about molecular and atomic spectroscopy due to this activity?

24. Explain how this activity helped you to improve your ability to find similarities and differences between two items (i.e., spectra).

Application

25. Many descriptions of microwave ovens explain that the molecules vibrate and heat up. The typical microwave works at a frequency of 2.45 gigahertz (10^9 s^{-1}). Does the microwave radiation increase the vibrational motion of the molecules? (Note: $\nu\lambda = c = 2.998 \times 10^8$ m/s)

26. a. Calculate the wavelength of light needed for an atomic absorbance process requiring 405 kJ/mol of energy. ($h = 6.626 \times 10^{-34}$ J s, $c = 2.998 \times 10^8$ m/s)

   b. In what region of the electromagnetic spectrum is this light found?

   c. Most atomic absorbance processes between atomic orbitals occur in the UV/Vis region of the electromagnetic spectrum. Does the wavelength you calculated seem reasonable?

27. Alkanes such as n-hexane are used as solvents in UV spectroscopy. The energy difference for the $\sigma \to \sigma^*$ transition is 886 kJ/mol. What wavelength does hexane absorb? Why is n-hexane a good choice as a solvent for UV/Vis spectroscopy?

28. Is the following transmission spectra atomic or molecular? Explain your reasoning.

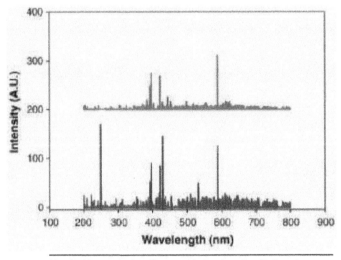

Used with permission: Spectrochimica Acta Part B: Atomic Spectroscopy, Volume 62, Issue 12, December 2007, Pages 1426-1432.

29. What are the differences in spectra between molecular and atomic absorption? Explain these differences using the type of transitions.

30. UV-Vis instruments used to measure molecular absorption typically have a broad wavelength light source while atomic absorption instruments use a light source that produces very narrow bands of light. Both are typically in the same UV-Vis wavelength range. Why don't they use the same source?

Appendix

UV-Vis Spectrum of Formaldehyde

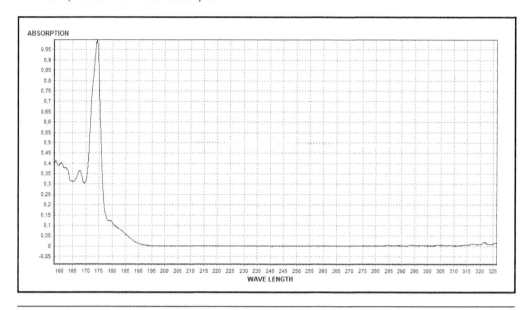

Used with permission: Chromalytica AB

Sodium Transmittance Spectrum

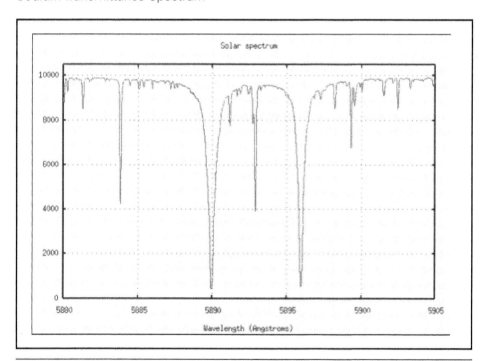

Used with permission: Michael Richmond, Rochester Institute of Technology

Introduction to Chromatography

Learning Objectives

Students should be able to:

Content

- Relate the separation process to features of the chromatogram, such as retention time and resolution.

- Explain how the chemical interactions between a solute and both the stationary phase and the mobile phase impact retention.

- Predict the elution order for a set of compounds given the mobile and stationary phase composition.

Process

- Relate molecular interactions and macroscale processes (Critical Thinking).

Prior knowledge

- An understanding of intermolecular forces at the general chemistry level.

Further Reading

- Harris, D.C. 2010. *Quantitative Chemical Analysis,* 8th Edition, W.H. Freeman: USA, Section 22-2, p.542-3.

- Skoog, D.A., F.J. Holler, S.R. Crouch, 2007. *Principles of Instrumental Analysis,* 6th Edition, Thomson/ Brooks-Cole, 26A-B, pp.763-768.

Authors

Caryl Fish, Mary Walczak, Ruth Riter and Paul Jackson

PGLANA004026a POGIL

Consider this...

In the experiment shown below in Figure 1, a glass column is packed with a solid material suspended in solvent. The solid material is called the stationary phase and the solvent is the mobile phase. Initially the same amounts of solutes A and B are added to the top of the column as shown in (a) below. As time progresses as shown in (b)-(e) solvent is allowed to flow through the column.

Figure 1 *Diagram showing a chromatographic separation of a mixture of solutes A and B.*

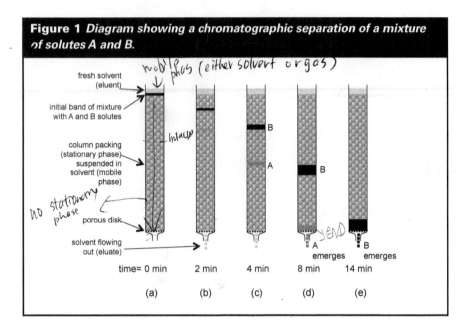

Key Questions

1. In Figure 1a, solutes A and B are mixed together in the initial band. As individuals, list differences observed as time progresses from (a)-(d). Compare your lists within the group. Develop a comprehensive list for the group.

The **retention time** (t_r) for a solute is the time needed after the mixture is placed on the column until that component emerges from the column. The retention factor (k') is the ratio of the time a solute spends in the stationary phase to the time spent in the mobile phase. Retention time for an unretained solute is the void time (t_m).

---

2. Using the elution times from Figure 1, what are the retention times for solutes A and B?

Consider this...

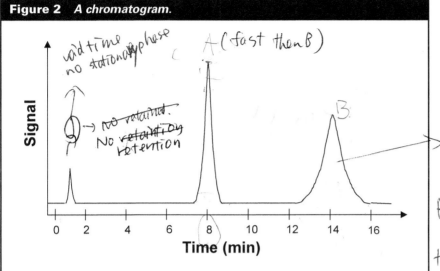

Figure 2 A chromatogram.

void time
no stationary phase
A (fast then B)
→ no retained.
No retention
B

Signal

Time (min)
0 2 4 6 8 10 12 14 16

The plot of the signal detected at the outflow of the column in Figure 1 during the elution of the solute mixture A and B. The first peak represents the time it takes for an unretained substance to pass through the column; it has no interaction with the stationary phase.

B is more big, broad
defuse is better
A & B has retentiime beaus
there add stationaty phase.

Key Questions

3. Which peak on the chromatogram corresponds to solute A and which peak corresponds to solute B? Label each peak as either A or B. Does your group agree with these labels?

4. The retention time determination should be consistent from person to person. From the chromatogram what part of the peak should be used to give the most consistent retention times? Explain your reasoning to your group.

The void time is a min.
↓ void time

5. Looking at the chromatogram, determine the void time for this column.

The void time is time trave. so we know
① is void time, ② is the A & ③ is B.

6. The void time is the same for all the solutes in this chromatogram. Discuss with your group why the void time is equal to the time the solutes spend in the mobile phase. Record your agreed upon explanation.

Because void tim is all equeal true to the A & B, A & B has stuck between.
stational phase, the void time is how much mobile phase past Throuyh.
fost (No retained.
reto A & B has retained time, A is fast then B, Boca A is 8 min B is 14 min

7. If the retention time of a solute is the time the solute spends in both the mobile phase and the stationary phase, how could you determine the time the solute spends in only the stationary phase?

If there is stationary phase, then solute is slow, not fost as
void timet   (or stack)
A is fast then B, beacue A is 8 min and B is 14 min.

8. Write an equation to show how the retention factors for each solute are determined from the chromatogram in Figure 2 and calculate the retention factors for solutes A and B. Note that the retention factor is a unitless number.

A = 8 min

B = 14 min

9. What is the relationship between the broadness of the peaks A and B in Figure 2 and their bandwidths shown in Figure 1? B is more broad than A, B is more defused. than A.

10. Describe the relationship between the retention factor and characteristics of the peaks on the chromatogram such as retention time and peak broadness.

B is biger then A. B is more defused.

In chromatography the term resolution (R_s) is used to express the quality of a separation between two peaks.

11. Look at Figure 2 and discuss with your group how changes in retention time and the broadness of peaks A and B would influence R_s. Write a summary of your group consensus below.

A is less broad than B.

12. Resolution can be computed by dividing the difference in retention times between the two peaks by their average width along the baseline.) If a value of R_s =1.50 represents baseline resolution between two peaks in time units, will the R_s value for the separation shown in Figure 2 be greater than or less than 1.50? Briefly explain your answer, and then verify it with a quick calculation by estimating these values.

because B is about 3.5 for width ⟩ average is it +1 = 4.5÷2
A is about 1 for width = 8.25

different in retention time = 14−8 = 6
average width

6/2.25 = 2.66 → grater than 1.50.

So because 2.25 is grater than 1.50, that means it is achieve baseline of resolution.

13. Based on your understanding of resolution, speculate as to how a chemist might improve the resolution of two chromatographic peaks.

Consider this...

Figure 3

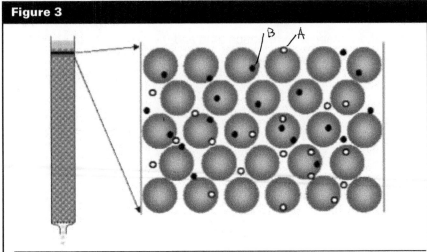

This is a snapshot microscopic view of individual solute molecules A (open circles) and B (black circles) in the vicinity of a solid stationary phase porous spherical particle (silica, $SiO2$ with a large number of surface Si-OH groups) in a column. Those solute molecules touching a stationary phase particle should be considered in the stationary phase. The mobile phase (hexane, C_6H_{14}) in which the solute is dissolved is not shown.

Key Questions

14. Examine the distribution of solutes A and B between the mobile phase and the stationary phase. Which solute has more molecules in the mobile phase?

15. This distribution of the solute molecules is similar to a partitioning between the stationary phase and the mobile phase. We can define a distribution constant (K) as the ratio of the concentration of the solute in phase 2 to the concentration in phase 1. In chromatography the mobile phase is considered phase 1 and the stationary phase is phase 2. Write the distribution constant expressions for solute A (K_A) and B (K_B).

16. Using the number of solute molecules in each phase in Figure 3, quickly approximate the distribution constants for solutes A and B.

17. If the interaction between the mobile phase and the solute is stronger than the interaction between the stationary phase and the solute, which phase do solute molecules partition preferentially into, the mobile or stationary phase? One group member should explain his/her reasoning to the group.

Keep in mind that Figure 3 is a snapshot as if all movement magically stopped in the column. Obviously, this does not occur during the separation process. Remember that this is a dynamic process and solute molecules are constantly moving between the mobile phase and the stationary phase. In addition, the process of separation is not an equilibrium process.

---

18. The solute will move through the column only if it is in which phase?

19. If a solute molecule partitions preferentially into the mobile phase, would you expect it to have a relatively long or short retention time? Do the relative values of K_A and K_B in KQ 16 support your conclusion? Explain.

20. Re-read the description of the mobile phase and stationary phase in Figure 3. Describe the mobile and stationary phases as polar or non-polar.

21. Discuss with your group whether solute A or B is the most polar, given that Solute B is retained longer on the column. Once you have reached a consensus, describe your group's reasoning below.

22. Discuss and explain in complete grammatically correct sentences your group's consensus of why Solute A elutes before solute B.

23. How confident are you in predicting the elution order of compounds in a separation? If you are not confident, what additional questions do you have?

24. What is one way this activity improved your ability to visualize the microscopic separation processes embedded in the appearance of a chromatogram?

Applications

25. In the experiment illustrated in Figures 1-3, which of these variables could change the elution order? Explain your reasoning.

Mobile phase composition   Length of column   Solvent flow rate

Solute composition   Stationary phase composition

26. Suppose you wish to decrease the amount of time for all the peaks to elute but continue to completely separate the compounds in Figure 2. Which of the following variables would affect this change? Explain your reasoning.

Mobile phase composition   Length of column   Solvent flow rate

Solute composition   Stationary phase composition

27. Which of the variables you identified in Q26 would be easiest to change and give the desired effect? Explain your reasoning.

28. Suppose you repeat the experiment in Figures 1 and 2 with the same mobile phase and stationary phase, but using methanol and cyclohexane as solutes. Which solute would elute first? Explain your reasoning.

29. Stationary phases come in different varieties. Which of the two analytes listed for each stationary phase has the greater retention time (or is retained more)? Explain your answer.

Stationary Phase	Retention mechanism	Analytes
Polar Solid	adsorption	H_2O and benzene
Nonpolar Liquid	solubility	H_2O and benzene
Ionic	electrostatic interaction	Na^+ and Mg^{2+}
Porous solid	pore penetration	ethylene and polyethylene

Band Broadening Effects in Chromatography

Learning Objectives

Students should be able to:

Content

- Explain how band broadening (i.e., band dispersion) is related to linear flow rate.

- Discuss how the three components in the van Deemter equation affect bandwidth.

Process

- Given a plot of data, identify the mathematical relationship (Critical thinking).

- Sketch graphs of known mathematical relationships (Information processing).

Prior knowledge

- Identify polar and nonpolar molecules and solvents.

- Intermolecular forces.

- Introduction to chromatography and chromatograms.

Further Reading

- D.C. Harris, *Quantitative Chemical Analysis*, 7th Edition, 2007 W.H. Freeman: USA, Section 23-2, p.516-521.

- D.A. Skoog, F.J. Holler, S.R. Crouch, *Principles of Instrumental Analysis*, 6th Edition, 2007, Thomson/ Brooks-Cole, 26A-B, pp.771-775.

- *The van Deemter Equation: A Three-Act Play* by Christa Colyer, Department of Chemistry, Wake Forest University, Winston-Salem, North Carolina. http://ublib.buffalo.edu/libraries/projects/cases/vandeemter/vandeemter.html

- S. J. Hawkes. "Modernization of the van Deemter Equation for Chromatographic Zone Dispersion." *J. Chem. Educ.* 1983, 60(5), pp.393-398.

- Patel, et al. *Anal. Chem.* 2004, 76, pp.5777-5786.

- Kirkland, J.J. *American Laboratory* April 2007, 18-21; Li & Carr. *Anal. Chem.* 1997, 69, pp.2193-2201.

- Mattice, John. "If You Were a Molecule in a Chromatography Column, What Would You See?" J. Chem. Educ. 2008, 85(7), pp.925-928.

Authors

Caryl Fish, Paul Jackson, Mary Walczak, Ruth Riter (Shepherd)

PGLANA004027a

Consider this...

Figure 1

The left illustration (A) shows a glass chromatography column packed with porous, spherical particles surrounded by solvent. Two solute bands are also shown, and the mobile phase is flowing from top to bottom of the column. The center drawing (B) is a microscopic view (not drawn to scale) of 14 particles in this column. Five (1-5) solute molecules are shown for simplicity. All the space between the stationary phase particles is filled with mobile phase which is not shown in the figure. On the right (C), a picture of a single, porous, spherical particle obtained with a scanning electron microscope (SEM) is shown.

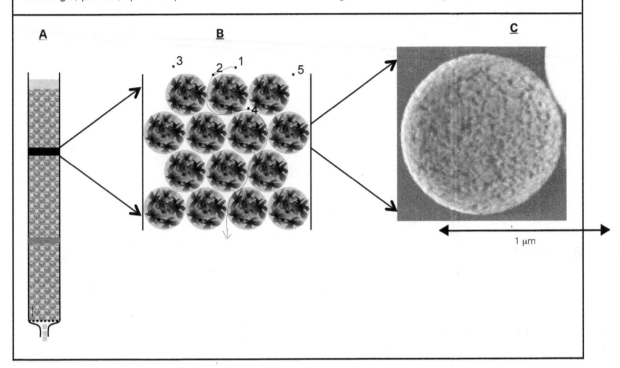

A   **B**   **C**

1 µm

Key Questions

Multiple Paths

1. Consider solute molecule 1 in Figure 1 (B). Each group member should independently draw a path for the molecule to move through this section of the column.

   mutipul   path = some take short, long. ways,

2. Compare your results. Which path is shortest? Which path is longest?

[1] Jacoby, Mitch. CE&N. 2008, 86(17), pp.17-23.

3. Considering the other molecules (2-5), draw possible paths for these molecules as they move through the figure. What effect would a large number of different solute paths have on band broadening?

4. Draw chromatograms for a solute eluting from a column with
A) a small number of multiple paths? B) many multiple paths?

5. As the mobile phase flow rate increases, will the *number* of pathways change as the mobile phase flow rate increases? Based on this, sketch in a graph the relationship between bandwidth and mobile phase linear flow rate, u_x.

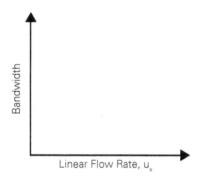

6. Write a proportionality between bandwidth and linear flow rate u_x. (*Hint:* bandwidth may be directly proportional, inversely proportional, or constant as u_x increases.)

Longitudinal Diffusion

7. Consider solute molecule 4 in Figure 1 (B). What is the most likely direction that this molecule will travel through the column while the mobile phase is flowing?

8. If we stop the mobile phase from flowing, solute molecule 4 will continue to move in the column due to diffusion. Draw arrows around this molecule to indicate the directions it might move.

9. What effect will diffusion of the solute molecules have on the broadness of the peaks eluting from the column?

10. Band broadening due to diffusion depends on flow rate. Will band broadening due to diffusion be more pronounced at slow or fast flow rates?

11. The relationship between the broadness of the peaks due to diffusion and mobile phase flow rate is represented by one of the following three graphs.

A) Choose which of the three models below represents this relationship and explain your rationale.

B) Write a proportionality between bandwidth due to diffusion and linear flow rate u_x.

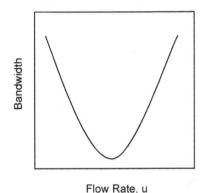 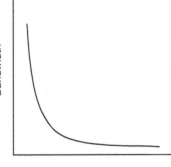  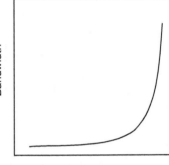

Mass Transfer

Chromatographic separation is a dynamic differential migration process. Solute molecules "distributed" or "dissolved" in the stationary-phase particles transfer to the mobile phase and vice versa continually throughout the column.

12. In Q1 and Q3 above, did your paths for molecules 1-5 include the molecules interacting with the porous stationary-phase particles? Yes __ / No__. Explain why the answer should be yes. If your answer is no, each group member choose one solute molecule in Figure 1 (B) and independently sketch a new path where the retention time is longer because the solute molecules interact with the stationary-phase particles.

13. Compare your drawn paths and briefly describe a difference and a similarity.

14. As the solute molecules move through the column in the mobile phase, **(a)** a portion of solute molecules spend little time interacting with the stationary phase and **(b)** a portion of solute molecules spend long periods of time in the stationary phase. Compare the progress of the **(a)** molecules down the column to the progress of the **(b)** molecules. Would band broadening increase or decrease due to the mass transfer of solute molecules to the stationary phase?

15. Now predict the progress of the **(a)** molecules if the flow rate of the mobile phase is increased.

16. Would you expect band broadening due to mass transfer to increase or decrease when the flow rate of the mobile phase increases? Explain your reasoning.

17. Assume that the relationship between bandwidth due to mass transfer and flow rate is linear. Write a proportionality between bandwidth and linear flow rate, u_x, and sketch a plot of this relationship.

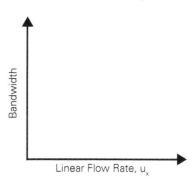

18. The moving of solute molecules from the mobile phase to the stationary phase and vice versa through the column is called mass transfer. Explain why and compare your group's answer with at least one other group in your class.

Consider this...

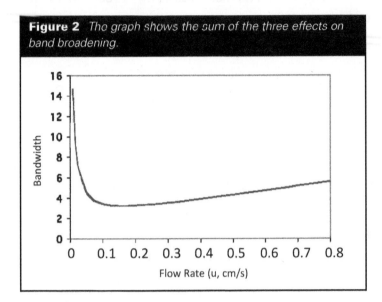

Figure 2 The graph shows the sum of the three effects on band broadening.

Bandwidth (y-axis)
Flow Rate (u, cm/s) (x-axis)

Key Questions

19. At flow rates < 0.05 cm/s, which of the three processes (multiple paths, mass transfer, or longitudinal diffusion) dominates band broadening?

20. At high flow rates (> 0.5 cm/s), which of the three processes is dominant?

21. Based on the three processes that lead to band broadening shown in Figure 2, what flow rate would you use to minimize band broadening?

22. Sketch on the plot above each of the three band broadening effects.

23. Would the bandwidth ever reach zero? Explain why or why not.

The relationship between band broadening and flow rate is known as the van Deemter equation where band broadening is defined in terms of "height equivalent to a theoretical plate", H:

$$H = A + (B/u) + Cu$$

The smaller the plate height, the narrower the bandwidth.

This equation is a useful qualitative model to describe how the three microscopic processes affect column dispersion.

---

24. Identify the three additive terms in the van Deemter equation using the relationships between the microscopic processes (multiple path, diffusion, and mass transfer) and mobile phase flow rate that you developed in Q6, Q11 and Q16.

25. After working through this activity and given the proportionality between an independent and dependent physical variables, how confident are you that you could sketch a graph

Applications

26. Predict how stationary-phase particle size would affect multiple paths band broadening mechanism and hence bandwidth.

27. In general, solute molecules in pores of the stationary-phase particles typically diffuse a distance equivalent to the particle diameter to leave the particle. Predict how stationary-phase particle size would affect mass transfer and hence bandwidth.

28. If the stationary-phase particle size in a column is decreased, does bandwidth increase, decrease, or stay the same? Explain your reasoning.

29. If the solute molecules and stationary phase are both nonpolar, how would increasing the polarity of a mobile phase affect bandwidth?

30. For gas chromatography, in an open tubular column where the stationary phase lines the walls of the column (as opposed to a packed column), A=0 in the van Deemter equation. Explain why this is true.

31. Increasing temperature of a chromatographic separation allows the flow rate of the mobile phase to be increased by a factor of 5 while maintaining comparable bandwidths. Explain how temperature affects mass transfer processes and hence bandwidth.

32. What impact (increase, decrease, or no change) does each of the following conditions have on the individual components of the van Deemter equation and consequently, band broadening?

	Multiple Paths	Diffusion	Mass Transfer
Increase temperature	No effect	increase	Decrease
Longer column	Packed column = increase Open Tubular column = No effect	No effect	No effect
Using a gas mobile phase instead of liquid	No effect	increase	Decrease
Smaller particle stationary phase	Decrease	Decrease (packing of particle hinders diffusion)	Decrease

$$H = H_{mp} + H_{Diff} + H_{mT} + H_{mT}$$

33. Below are experimentally determined van Deemter plots of column efficiency, *H,* vs. flow rate. *H* is a quantitative measurement of band broadening. The left plot is for a liquid chromatography application and the right is for gas chromatography. Compare and contrast these two plots in terms of the three band broadening mechanisms presented in this activity. How are they similar? How do they differ? Justify your answers.[2]

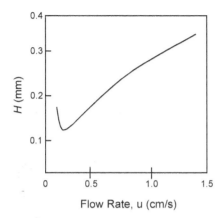 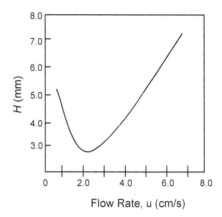

34. Figure 3 shows Van Deemter plots for a solute molecule using different column inner diameters (i.d.).

A) Predict whether decreasing the column inner diameters increase or decrease bandwidth.

B) Predict which van Deemter equation coefficient (A, B, or C) has the greatest effect on increasing or decreasing bandwidth as a function of i.d. and justify your answer.

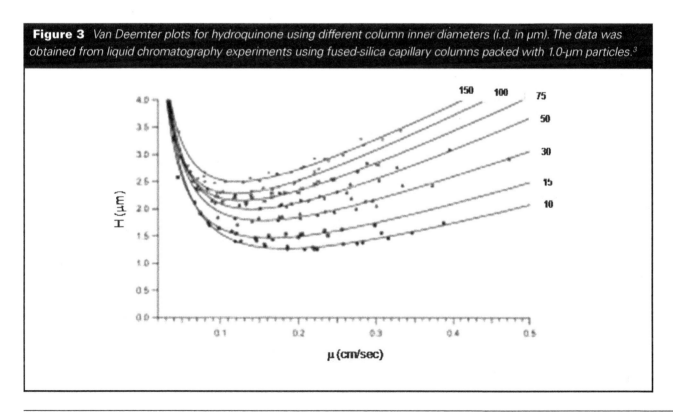

Figure 3 *Van Deemter plots for hydroquinone using different column inner diameters (i.d. in μm). The data was obtained from liquid chromatography experiments using fused-silica capillary columns packed with 1.0-μm particles.*[3]

---

[2] Plots are adapted from Figure 30-13 in Skoog, D. A., S. M. West, F. J. Holler, and S. R. Crouch. *Fundamentals of Analytical Chemistry,* 8th ed., Belmont, Ca:Brooks/Cole, 2004.

[3] Patel, et al. Anal. Chem. 2004, 76, pp. 5777-5786

---

35. Figure 4 shows scanning electron microscope (SEM) images of extruded sections of packing bed for two capillary columns of different diameters, **a)** 750 (bottom image) and **b)** 30-μm-i.d. Both columns are packed with the same stationary phase, spherical particles with 1-μm diameter.

A) When the columns were prepared, the figure shows that the column with the larger diameter has more packing irregularities. Explain this observation.

B) Predict what affect this should have on band broadening and discuss your prediction using the van Deemter terms.

C) Does this figure support your explanations in application question 33? Explain why or why not and make any changes in your answers in light of this figure.

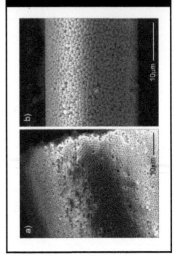

Figure 4 *SEM images of sections of packed columns for a) 750 and b) 30-μm-i.d. capillary columns.*[3]

Gas Chromatography or HPLC, Which Do You Choose?

Learning Objectives

Students should be able to:

Content

- Describe the differences and similarities between Gas Chromatography (GC) and High Performance Liquid Chromatography (HPLC).

- Compare sample preparation for GC and HPLC.

- Compare resolution, detection limit, and analysis time for a particular chromatographic separation.

Process

- Interpreting schematic diagrams (Information Processing).

- Identifying similarities and differences (Critical Thinking).

- Integrating prior knowledge to choose appropriate instrument (Critical Thinking).

Prior knowledge

- Introduction to chromatography.

- Band Broadening.

- Gas chromatographic and HPLC instrument details.

Further Discussion

- Instrument components including advantages and disadvantages of different component types.

- Determining concentrations from chromatographic data.

Further Reading

- D.C. Harris, *Quantitative Chemical Analysis*, 7th Edition, 2007 W.H. Freeman: USA, Chapters 24-25, pp.528-583.

- Gratz, L.D. et al. *J. Hazardous Materials,* 74 (2000) 37-46.

Author

Caryl Fish

PGLANA004028a

Consider this...

Figure 1 *Diagram of a GC instrument*

Typical GC configuration: Helium as the carrier gas, injector heated to 250° C, 1 μl of sample is injected, the column is a 30-meter, open tubular column, the column oven can range from 60° C to 300° C, the detector is a flame ionization detector that responds to most organic compounds, the output is a computer system that can record the retention time and area of each peak, but does not give structural information about the analytes.

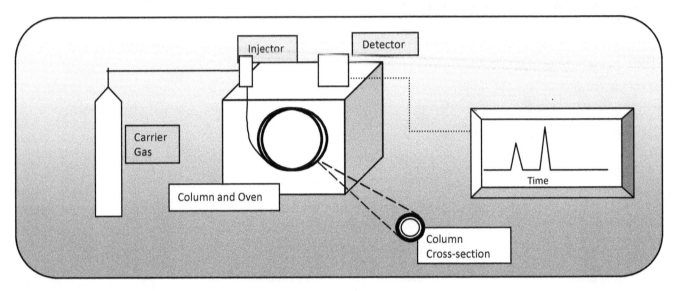

Figure 2 *Diagram of an HPLC instrument*

Typical HPLC System: Solvent is a mixture of water and a polar organic solvent, the ratio of solvents can be changed, the pump flows at 1-2 μl/min, 10–25 μl of sample are injected, an analytical column is made of non-polar organic compounds bonded to 5–10 um particles packed into a 20 cm long column, the detector is a ultraviolet detector that records absorbance at a particular wavelength, the output is a computer system that can record the retention time and area of each peak, but does not give structural information about the analytes.

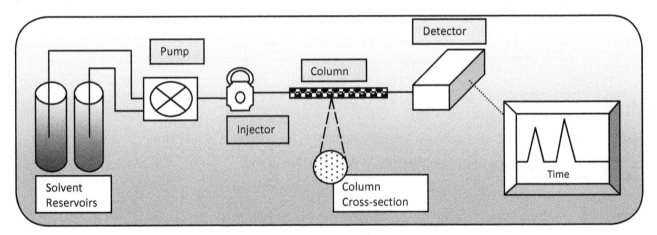

Key Questions

1. Determine three similarities and three differences between the GC and HPLC instruments. Compare your answers with your group and make a combined list.

2. Focus on the columns and examine the columns in the GC and HPLC. As a group, determine four differences.

3. Since the column is the heart of the separation process, describe how the differences in the GC and HPLC columns could impact the ability to separate compounds.

4. What is the mobile phase in the GC? In the HPLC?

5. Based on these mobile phases, what characteristic(s) must an analyte possess to be able to use a GC for analysis? An HPLC for analysis?

Typically organic compounds analyzed by GC must be volatile (with boiling points less than 500° C) and thermally stable (do not decompose at high temperatures). For reverse phase HPLC (polar solvent, non-polar stationary phase), the analytes must be soluble in polar solvents.

Consider this...

Sample preparation for GC and HPLC can have four purposes: **1)** put the analytes into the appropriate solvent for the instrument; **2)** adjust the concentration to fit the concentration range of the instrument; **3)** remove contaminants that could interfere with the analysis, **4)** chemically react the analytes to change their properties so they are more amenable to separation and detection (derivatization). For example, amino acids do not fluoresce. If they are reacted with 9-fluorenylmethyl chloroformate, the products fluoresce and can be detected by fluorescence on an HPLC.

Sample Prep Schemes

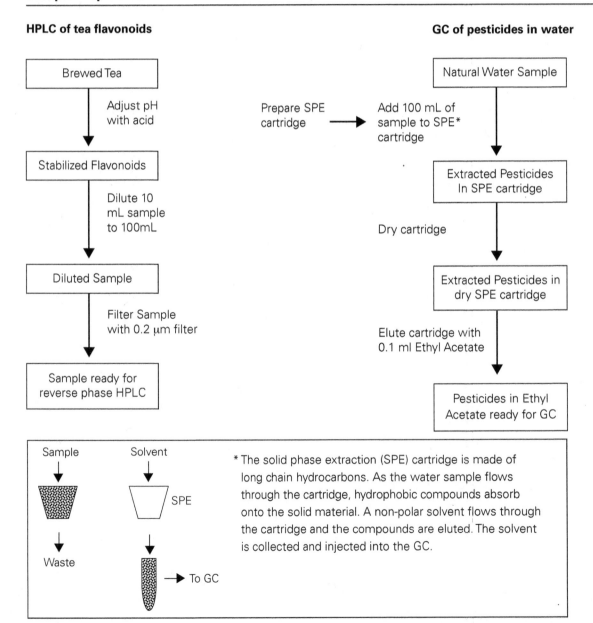

HPLC of tea flavonoids

Brewed Tea

↓ Adjust pH with acid

Stabilized Flavonoids

↓ Dilute 10 mL sample to 100mL

Diluted Sample

↓ Filter Sample with 0.2 μm filter

Sample ready for reverse phase HPLC

GC of pesticides in water

Natural Water Sample

Prepare SPE cartridge → Add 100 mL of sample to SPE* cartridge

↓

Extracted Pesticides In SPE cartridge

↓ Dry cartridge

Extracted Pesticides in dry SPE cartridge

↓ Elute cartridge with 0.1 ml Ethyl Acetate

Pesticides in Ethyl Acetate ready for GC

Sample   Solvent

↓   ↓

         SPE

↓   ↓

Waste

→ To GC

* The solid phase extraction (SPE) cartridge is made of long chain hydrocarbons. As the water sample flows through the cartridge, hydrophobic compounds absorb onto the solid material. A non-polar solvent flows through the cartridge and the compounds are eluted. The solvent is collected and injected into the GC.

6. Examine the sample preparation scheme for the HPLC of flavonoids. This scheme addresses two of the four purposes of sample prep. Describe which purposes are addressed and how it addresses these purposes.

7. Examine the sample prep scheme for the pesticides in water. Describe the change in concentration of the pesticides from the original water sample to the GC ready sample.

8. Examine the HPLC sample prep scheme and determine the change in concentration for this particular type of sample.

9. How would you determine whether a sample requires dilution or concentration?

10. What would happen to polar contaminants in the water sample during the SPE extraction process?

11. Both of these schemes involve samples in aqueous solutions. What purpose is necessary in the GC sample prep that is not necessary for reverse-phase HPLC sample prep?

12. Discuss with your group what three purposes are addressed in the GC sample prep scheme.

13. The fourth purpose was not really addressed in either of these sample prep schemes. What properties of the analyte would make it difficult to analyze by GC (refer to Q5) and therefore require derivitization?

14. It is obvious the type of sample preparation necessary depends on the analytes and well as the sample matrix. Discuss with your group a set of guidelines that would determine the sample preparation necessary for both GC and reverse-phase HPLC. Record your group's consensus.

Consider this...

Polynuclear aromatic hydrocarbons (PAH) are formed during combustion of fossil fuels, refuse burning, coke ovens, and even smoked foods. Several PAHs are carcinogens or potential carcinogens and therefore monitored in the environment. Shown in Figure 2 are chromatograms of a standard containing the 16 EPA controlled PAHs. Figure 2A is a capillary column GC chromatogram while Figure 2B shows an HPLC chromatogram. For both analyses the instrumental parameters are shown on the chromatogram.[1]

GC

%

Figure 2A Gas chromatogram

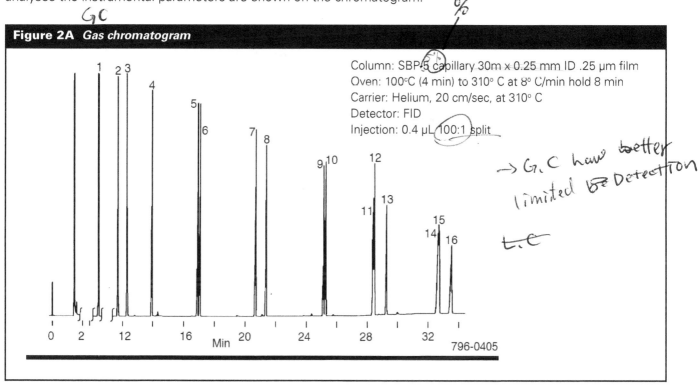

Column: SBP-5 capillary 30m x 0.25 mm ID .25 µm film
Oven: 100°C (4 min) to 310° C at 8° C/min hold 8 min
Carrier: Helium, 20 cm/sec, at 310° C
Detector: FID
Injection: 0.4 µL 100:1 split

→ G.C have better limited be Detection

t.C

796-0405

Figure 2B HPLC chromatogram

Column: C18 3.3 cm x 4.6 mm ID, 1.5 µm particles
Mobile phase: 35%-55% acetonitrile in water, 4 min gradient
Detector: Fluorescence
Flow rate: 1 mL/min, 230 bar
Injection: 10 µL

better resolution then GC.

LC has better stationary phase

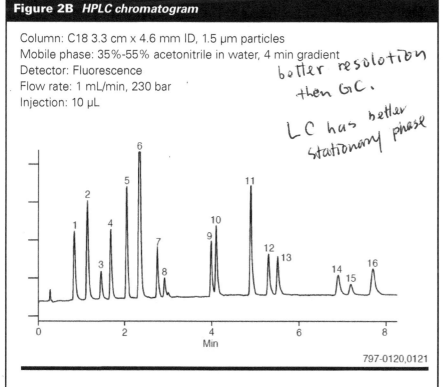

797-0120,0121

Concentration of PAHs (µg/mL)

		GC	HPLC
1.	Naphthalene	1000	500
2.	Acenaphthylene	1000	500
3.	Acenaphthene	1000	1000
4.	Fluorine	1000	100
5.	Phenanthrene	1000	40
6.	Anthracene	1000	20
7.	Fluoranthene	1000	50
8.	Pyrene	1000	100
9.	Benzo(a)anthracene	1000	50
10.	Chrysene	1000	50
11.	Benzo(b)fluoranthene	1000	20
12.	Benzo(k)fluoroanthene	1000	20
13.	Benzo(a)pyrene	1000	50
14.	Dibenzo(a,h)anthracene	1000	200
15.	Benso(g,h,i)perylene	1000	80
16.	Indeno(1,2,3-c,d)pyrene	1000	50

[1] Supelco, Application notes 108, 138, 1997.

Key Questions

15. Examine the instrument parameters used to produce each chromatogram. Fill in the table below for some of these parameters.

Instrument Parameters	GC	HPLC
Column	SBP-5	C18
Mobile Phase		35-55 % aceton in water
Detector	FID	Fluorence 50
Temp Ramp or gradient		4 min
Concentration of acenaphthene	1000	1000

The following questions will examine three important criteria for chromatographic analysis: resolution, retention time, and sensitivity/detection limits.

Resolution

16. Which of the instruments provides better resolution of peaks for the PAHs analyzed? Discuss your choice with your group and defend your choice using specific examples.

17. What pairs of peaks are particularly difficult to separate? Examine the table of PAH compounds at the end of this activity and describe why these pairs are problematic.

> **The C-18 column used in this HPLC analysis is a long chain hydrocarbon stationary phase. The SPB-5 is a non-polar a general purpose column that provides an elution order based on boiling point.**

18. Consider how the compounds interact with the stationary phase in both GC and HPLC. How does this explain why HPLC gives better separation for the PAH compounds?

Retention time

19. Compare the retention times of the last compound to elute from the GC and HPLC. Why is this time important?

20. What are the advantages to a chemist of a short total analysis time?

Sensitivity/Detection Limit

While information concerning sensitivity and detection limits is not given for these analyses, we can get a sense of these parameters by comparing the mass of the compounds that were analyzed and the peaks on the chromatograms.

21. Using the concentration and the volume injected determine the mass of acenaphthene injected for each instrument.

22. In the GC analysis, a 100:1 split is used in the injector. This means that a portion of the vapor in the injector is swept away and only 1/100 of this vapor travels into the column. This is essentially a 100 times dilution. Given this information what is the mass of the acenaphthene in the column?

23. Now compare the peak heights and areas of the acenaphthene in the GC and HPLC chromatograms. For these particular analyses, which would you infer has the lower detection limits?

24. Considering the three criteria examined here (resolution, retention time, and detection limit), discuss with your group which of these two methods you would use to analyze PAHs in a wastewater sample. Record your group's choice and reasoning.

25. What other criteria could be considered when choosing an instrumental method?

26. Describe two examples of how your understanding of the differences between GC and HPLC has improved.

27. How are you more adept at interpreting schematic diagrams?

Applications

28. Highlight the differences between HPLC and GC instruments.

29. Since both HPLC and GC separate mixtures of compounds, why would a laboratory (i.e. a forensics lab) need both instruments?

30. Develop a checklist or set of questions about a sample to help determine if it would be best run on a GC or an HPLC.

31. Based on your understanding of instrument components (columns, detectors, injectors, etc) what could be changed in both the GC and HPLC to improve the analysis (resolution, retention times, or sensitivity) of PAHs?

32. Examine the following table of compounds and determine which might be suitable for GC analysis and which for HPLC (Some could be both).

Compound	Molar Mass (g/mol)	Boiling Point (°C)	Stability	Solubility	Instrument?
n-Butanol	130.23	195	Stable	Alcohol, ether, acetone	
Chlorobenzene	112.56	131	Stable	Alcohol, ether, benzene	
Pyrene	202.26	393	Stable	Benzene	
Citric Acid	192.12	—	Decomposes	Water, alcohol	
Vitamin D	384.64	—	Decomposes	Alcohol, ether	

33. The flame ionization detector (FID) in GC and the UV detector in HPLC both respond to many organic compounds and therefore can be used for a variety of mixtures. They do not, however, give any structural information about the compounds. In some cases different compounds can have the same retention time, therefore retention time alone is not an absolute identifier of a chemical compound. Using additional resources, find 3 different detectors for GC and/or HPLC that can be used to identify the compounds after separation and discuss the advantages of each of these detectors.

Structures of Polyaromatic Hydrocarbons[2]

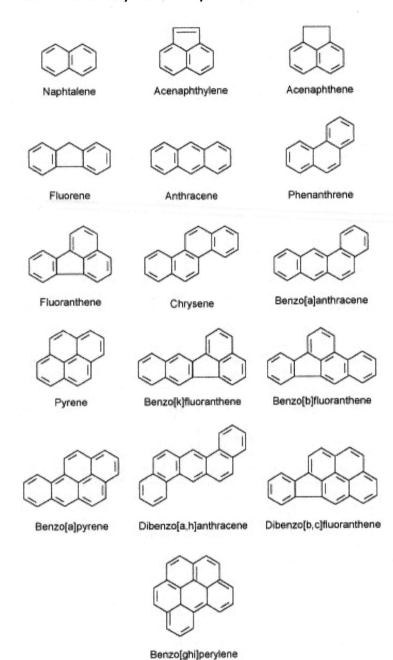

Naphtalene

Acenaphthylene

Acenaphthene

Fluorene

Anthracene

Phenanthrene

Fluoranthene

Chrysene

Benzo[a]anthracene

Pyrene

Benzo[k]fluoranthene

Benzo[b]fluoranthene

Benzo[a]pyrene

Dibenzo[a,h]anthracene

Dibenzo[b,c]fluoranthene

Benzo[ghi]perylene

[2] Journal of Chromatography A, vol. 885, issues1-2, 14 July 2000, p.273-290.

ANA-POGIL Project Activities: Topics and Learning Goals

Analytical Tools		Quant	Instrumental	Environmental	Bio-analytical
Content Goals	**Process Skills**				
Accuracy, Precision and Tolerance: Sorting Out Glassware		✓		✓	✓
• Differentiate between accuracy and precision. • Compare the tolerance of various pieces of glassware. • Classify data in terms of accuracy and precision	• Interpret tables of volumetric glassware tolerances (Information processing) • Infer based on tabulated data (Critical thinking) • Collaborate with group members (Teamwork)				
Solutions and Dilutions		✓		✓	✓
• Design a procedure for making a particular solution and assess the advantages of different approaches • Choose the appropriate glassware to ensure the desired level of precision of a particular solution • Convert between different concentration units (e.g., ppm to M)	• Develop alternative pathways for diluting solutions (Information processing) • Design approaches for preparing solutions (Problem solving) • Infer chemical processes based on reactions (Critical thinking)				
Classical Analytical Methods – A Design Perspective On Volumetric Measurement		✓			
• Describe the characteristics of the chemical substances and reactions used in volumetric analysis, particularly as they affect accuracy and precision. • Explain the rationale for the procedural steps in volumetric analysis, particularly as they affect accuracy and precision. • Explain how visual indicators are used • Explain the role of primary standard materials.	• Apply chemical concepts and knowledge of mass and volume measurement to making design decisions about volumetric measurement (Critical thinking) • Use typical titration measurement calculations (Problem solving)				
Sample Preparation		✓	✓	✓	
• Evaluate sample preparation methods for completeness. • Determine the concentration of analyte in the sample from the concentration in the digestate. • Explain how to assure sample integrity with respect to contamination and accuracy.	• Form an understanding of the accuracy of a measurement (Critical thinking) • Analyze methods for robustness and sources of error, and overall quality (critical thinking)				
Instrumental Calibration		✓	✓	✓	✓
• Use a correlation coefficient to ascertain the fit of data to a calibration curve • Determine the minimum detectable concentration and lower limit of quantitation and distinguish between these. • Determine the linear dynamic range from a set of calibration data.	• Interpret tabulated data (Information processing) • Produce an xy graph of data (Graphing) • Interpret calibration curves (Information Processing)				

Analytical Tools	Quant	Instrumental	Environmental	Bio-analytical

Quality Assurance Measures – How Good is the Data?

	Quant	Instrumental	Environmental	Bio-analytical
	✓	✓	✓	

- Calculate the spike recovery and relative deviation between replicate samples.
- Use spike recovery and relative deviation between replicate samples to assess and alidate sample analysis quality.
- Use spike recovery and relative deviation between replicate samples to assure accuracy and precision of an analysis.

- Validate data (Problem solving)
- Interpret tabulated data (Information processing)

Instrumental Calibration: Method of Standard Addition

	Quant	Instrumental	Environmental	Bio-analytical
	✓	✓		

- Recognize when the method of standard addition is needed (i.e. matrix effects)
- Distinguish between the method of standard additions and normal calibration using standards.
- Determine the concentration of an unknown in an original sample using the method of standard additions.

- Interpret calibration graphs (Information processing)
- Calculate concentrations of unknown solutions from standard addition data (Information processing)
- Read and interpret a standard method (Information processing)

Instrumental Calibration: Method of Internal Standards

	Quant	Instrumental	Environmental	Bio-analytical
	✓	✓	✓	

- Describe calibration by internal standards as a method to deal with irreproducible instrumental results
- Calculate the concentration of an analyte in a sample when analyzed using the method of internal standards.
- Determine the need for the internal standard method.

- Construct and interpret graphs. (Information processing)

Interlaboratory Comparisons

	Quant	Instrumental	Environmental	Bio-analytical
	✓	✓	✓	

- Explain how random and systematic error affects experimental results.
- Describe what the purpose of an interlaboratory study is and how it might be conducted.
- Explain how spike recovery may be used to evaluate accuracy of a method and of a laboratory's performance.

- Draw inferences from the data presented in graphical form (Information processing)

Sampling Error and Sampling Designs

	Quant	Instrumental	Environmental	Bio-analytical
	✓	✓	✓	

- Describe a sample as a part taken from the whole system of interest, where the part is representative of the chemical composition of the whole.
- Describe the different types of sampling designs and how they address compositional or temporal heterogeneity in the system of interest.
- Estimate the relative importance of sampling error relative to measurement error.

- Seek patterns in tabulated data and summarize trends (Information processing)
- Design a sampling strategy for a practical case (Problem solving)

Statistics		Quant	Instrumental	Environmental	Bio-analytical
Content Goals	**Process Skills**				
Errors in Measurements and Their Effect on Data Sets		✓			
• Classify different sources of experimental error and determine their effect on the mean and dispersion of the data. • Develop the ability to judge whether an analytical balance has been used correctly. • Define situations in which it is reasonable to reject data.	• Develop critical skills in assessing the quality of data (Critical thinking) • Identify qualitative difference among data sets (Information processing) • Develop generalizations from tabulated data (Information Processing)				
The Gaussian Distribution		✓	✓		
• Explain how the shape and position of a Gaussian Distribution are controlled by the mean and standard deviation. • Explain how histograms can be used to illustrate the frequency and distribution of data, and how the sample size affects the accuracy of the histogram. • Describe the relationship between the Gaussian distribution function and probability.	• Interpret graphs of data, especially histograms. (Information processing) • Estimate the distribution parameters of a data set (Information processing) • Draw conclusions about data sets. (Critical thinking)				
Statistical Tests of data: The t Test		✓		✓	
• Explain how the t value changes due to changes in the means, the standard deviations, and the number of data. • Explain how the t test is used to make decisions regarding significance, and be able to use Excel's t test output to make such decisions. • Develop an understanding of the t value's relationship to the probability of a significant difference.	• Interpret tables of data. (Information processing) • Draw conclusions about data sets. (Critical thinking) • Express concepts in grammatically correct sentences. (Communication)				
Statistical Tests of Data: The F Test		✓			
• Explain how the F value changes with respect to the standard deviations of the data sets. • Explain how the F test is used to make decisions regarding significance, and be able to use Excel's F test output to make such decisions. • Explain how the results of the F test will affect the t test.	• Interpret tables of data. (Information processing) • Draw conclusions about data sets. (Critical thinking) • Express concepts in grammatically correct sentences. (Communication)				
Linear Regression for Calibration of Instruments		✓	✓	✓	
• Describe linear regression as an averaging process • Interpret the results of a linear regression analysis • Evaluate the quality of a linear model for calibration of an instrumental measurement	• Construct and interpret graphs (Information processing)				

Equilibrium		Quant	Instrumental	Environmental	Bio-analytical
Content Goals	**Process Skills**				
The Importance of Ionic Strength		✓			
<ul><li>Be able to describe the role of electrolytes in determining the ionic strength of solutions.</li><li>Be able to explain, on a microscopic level, the effect of ionic strength on equilibrium concentrations.</li><li>Be able to calculate the ionic strength of a solution and predict its impact on equilibrium concentrations.</li></ul>	<ul><li>Identifying key properties of electrolyte solutions (Problem solving)</li><li>Depicting a shared understanding of microscopic solution interactions (Information processing, teamwork)</li><li>Challenging assumptions (Critical thinking)</li></ul>				
The Importance of Ionic Strength – Biochemistry Focus					✓
<ul><li>Use the concept of ionic atmosphere to explain how ions behave in solution.</li><li>Be able to calculate the ionic strength of a buffer solution.</li><li>Compare the relationship between ionic strength and concentration for various buffers.</li></ul>	<ul><li>Identify key properties of electrolyte solutions (Problem solving)</li><li>Form a shared understanding of microscopic solution interactions (Teamwork)</li><li>Challenging assumptions (Critical thinking)</li></ul>				
Activity and Activity Coefficients		✓			
<ul><li>Describe the impact of ionic strength on activity coefficients and the equilibrium concentrations of slightly soluble salts.</li><li>Develop strategies for calculating activity coefficients, activities and equilibrium concentrations in various electrolyte solutions.</li><li>Discern the ionic strength range for which activities need to be used in equilibrium calculations.</li></ul>	<ul><li>Finding, predicting trends in data (Information Processing)</li><li>Identifying similarities and differences in solutions (Critical thinking)</li><li>Estimating values, choosing solving strategies for various concentration regions (Problem solving)</li></ul>				
Multiple Equilibria: When Reactions Compete		✓		✓	
<ul><li>Explain how a competing equilibrium affects the solubility of a sparingly soluble compound in water.</li><li>Predict the pH dependence of the solubility of a sparingly soluble salt.</li></ul>	<ul><li>Identifying key reactions (Problem solving)</li><li>Challenging assumptions (Critical thinking)</li><li>Predicting pH dependence (Information processing)</li></ul>				
pH of Solutions of Strong Acids and Bases		✓		✓	
<ul><li>Calculate the pH of solutions containing any concentration of strong acid or base.</li></ul>	<ul><li>Recognize when the result of a calculation makes sense (Critical thinking)</li><li>Identify assumptions (Problem solving)</li></ul>				
Acid-base Distribution Plots		✓		✓	
<ul><li>Describe how the molar concentrations of mono- and polyprotic weak acids and their conjugate bases vary with pH.</li><li>Identify the principal species resulting from the dissociation of a weak acid at a given pH.</li></ul>	<ul><li>Sketch ionic distribution graphs given acid-base parameters (Critical thinking, visualization)</li><li>Interpreting graphs (Information processing)</li></ul>				

Equilibrium

Content Goals	Process Skills	Quant	Instrumental	Environmental	Bio-analytical
Acid-Base Distribution Plots – Biochemistry Focus					✓
• Describe how the molar concentrations of mono- and polyprotic weak acids and their conjugate bases vary with pH. • Use an acid-base distribution plot to determine the pKa values of weak acids and to determine the appropriate buffering region for a weak acid/conjugate base system. • Use the pKa value of the functional groups of amino acids to estimate a value for the pI of an amino acid.	• Sketch ionic distribution graphs given acid-base parameters (critical thinking– visualization) • Interpret graphs (information processing)				
The Acid-Base Distribution Functions		✓			
• Explain what information acid-base fractional distribution functions provide. • Use acid-base distribution functions to calculate the concentration of a species arising from any weak acid at a given pH.	• Extend the application of a method to new situations using pattern recognition (Critical thinking) • Work effectively with others to reach a consensus (Teamwork) • Evaluate answers for reasonableness (Critical thinking)				
The Buffer Zone: What is a Buffer and in What pH Range is it Effective?		✓		✓	
• Explain how the concentration of buffer components determines the pH of the solution. • Explain the effects of additions of strong acids/bases to buffered and non-buffered solutions • Determine the relationship between the pKa of an acid and the optimal pH range of a buffer in order to select an appropriate buffer for a particular pH range.	• Interpret tabulated information (Information processing) • Recognize and predict trends in data (Critical thinking) • Generalize problem solutions (Problem solving and critical thinking) • Include all group members (Teamwork)				
The Buffer Zone — Biochemistry Focus					✓
• Explain how the pH of a solution is altered by the concentration of buffer components. • Determine the relationship between the pKa of an acid and the optimal pH range of a buffer in order to select an appropriate buffer for a particular pH range. • Calculate the buffer capacity of a solution.	• Interpret tabulated information (Information processing) • Recognize and predict trends in data (Critical thinking) • Generalize problem solutions (Problem solving and critical thinking) • Include all group members (Teamwork)				
When Acids and Bases React: Laboratory		✓			✓
• For a given acid/base reaction, distinguish between equivalence point and endpoint. • Given a titration curve, identify whether a monoprotic acid is "strong" or "weak". • Given a titration curve, identify the main features of the curve, i.e. (1) whether an analyte is an acid or base, (2) whether an analyte is a mono- or polyprotic acid, and (3) what is/are the analyte's pK value(s).	• Obtain titration curve using pH probe, drop counter, and data acquisition software (Experimental technique) • Determine the equivalence point of a titration curve from 1st and 2nd derivative plots of the curve using computer software, i.e. Excel, Vernier software, etc. (Information processing) • Predict the appearance of a titration curve (critical thinking)				

Electrochemistry

Content Goals	Process Skills	Quant	Instrumental	Environmental	Bio-analytical
Electrochemistry: The Microscopic View of Electrochemistry		✓			
• Describe microscopic-level processes occurring at electrodes and explain how these processes result in formation of the electrochemical double layer. • Explain the origins of electrochemical potential. • Use E° values to predict the favored reaction in electrochemical reactions.	• Interpret pictures of electrochemical processes. (Information Processing) • Draw and interpret microscopic pictures of electrochemical processes. (Information Processing) • Draw and interpret graphical representations of electrochemical processes. (Critical Thinking)				
Electrochemistry: Calculating Cell Potentials		✓			
• Apply the criteria for an electrochemical cell in the standard state to determine if a cell is initially in the standard state. • Predict how the ion concentrations in an electrochemical cell will change as an electrochemical cell circuit is completed and allowed to run. • Determine Ecell for a given electrochemical cell and decide whether the reaction is spontaneous in the forward or reverse direction	• Interpret drawings of electrochemical cells (Information processing) • Compare and contrast different electrochemical cells. (Critical thinking)				

Spectrometry

Content Goals	Process Skills	Quant	Instrumental	Environmental	Bio-analytical
The Beer-Lambert Law		✓	✓	✓	
• Identify each term in the Beer-Lambert law and explain its effect on absorbance. • Explain the relationship between transmittance and absorbance. • Use Beer's Law in quantitative measurements.	• Prepare and interpret graphs (Information processing) • Develop mathematical expressions to describe data (Problem solving).				
The Beer-Lambert Law – Biochemistry Focus					✓
• Identify each term in the Beer-Lambert law and explain its effect on absorbance. • Explain the relationship between transmittance and absorbance. • Use Beer's Law in quantitative measurements.	• Prepare and interpret graphs (Information processing) • Develop mathematical expressions to describe data (Problem solving).				
Atomic and Molecular Absorption Processes		✓	✓		
• Elate the wavelength and energy of light absorbed to molecular and atomic transitions. • Compare molecular and atomic absorption processes and explain differences in bandwidth in the resulting spectra.	• Compare and contrast spectra and absorption processes (Critical thinking)				

Introduction to Spectrometers			✓		
• Identify the components of a UV-Vis spectrophotometer. • Describe characteristics of the source, wavelength selector, sample holder, and detector in a UV-Vis spectrophotometer. • Compare different types of instrument components based on these key characteristics and the goals of the instrumental analysis.	• Interpret graphical data. (Information processing) • Interpret schematic diagrams and relate to real instruments. (Information processing) • Critical thinking by comparison and contrast.				
Infrared Spectroscopy: Vibrational Modes			✓		
• Classify a vibration as a bend or a stretch. • Predict the number and type of motions for a particular molecule. • Explain the relationship between Infrared (IR) absorbance and changes in dipole moment.	• Use the chemical education digital library to view molecular vibrations and corresponding IR spectra • Infer general relationships from specific examples (Generalizing) • Specify the frequencies of some infrared vibrational modes (Integrating prior knowledge)				
Mass Spectrometry - Interpretation			✓		
• Explain the source of fragment ions in electron impact mass spectra. • Identify the base peak and parent (molecular ion) in a mass spectrum. • Recognize patterns in the mass spectra and use them to identify functional groups.	• Comparing and contrasting graphical data (critical thinking) • Analyze graphical information (Information processing)				

Chromatography and Separation

Intro to chromatography		✓	✓	✓	
• Relate the separation process to features of the chromatogram, such as retention time, and resolution. • Explain how the chemical interactions between a solute and both the stationary phase and the mobile phase impact retention. • Predict the elution order for a set of compounds given the mobile and stationary phase composition.	• Relate molecular interactions and macroscale processes (Critical thinking)				
Band Broadening Effects in Chromatography		✓	✓		
• Explain how band broadening (i.e band dispersion) is related to linear flow rate. • Discuss how the three components in the van Deemter equation affect bandwidth.	• Given a plot of data, identify the mathematical relationship (Critical thinking) • Sketch graphs of known mathematical relationships (Information processing)				
Gas Chromatography or HPLC, Which do You Choose?		✓	✓	✓	
• Describe the differences and similarities between Gas Chromatography (GC) and High Performance Liquid Chromatography (HPLC). • Compare sample preparation for GC and HPLC. • Compare resolution, sensitivity, and analysis time for a particular chromatographic separation.	• Interpreting schematic diagrams (Information Processing) • Identifying similarities and differences (Critical thinking) • Integrating prior knowledge to choose appropriate instrument (Critical Thinking)				

Introduction to Gel Filtration Chromatography		✓	✓		✓
• Explain how the sample protein characteristics and the characteristics of both the stationary phase and the mobile phase impact elution. • Use their understanding of how proteins interact with gel filtration resins to derive an expression for the term kav.	• Relate molecular interactions and macroscale processes (Critical thinking) • Derive mathematical expressions that describe the separation process. (Critical thinking)				

The ANA-POGIL project is funded by a four year NSF grant (DUE # 0717492) entitled: "A New Approach to Analytical Chemistry: The Development of Process Oriented Guided Inquiry Learning Materials."
The Principal Investigator is Juliette Lantz (Drew University) and the Co-Principal Investigator is Renee Cole (University of Iowa).

ANA-POGIL Activities & Analytical Texts Mapping

Activity	Harris Chapter(s) 8th ed.	Skoog Chapter 6th ed.	Christian Chapter 9th ed.
Analytical Tools			
Accuracy, Precision, and Tolerance: Sorting Out Glassware	2 (Tools of the Trade) 3 (Experimental Error)	Appendix 1	3 (Statistics and Data Handling in Analytical Chemistry)
Solutions and Dilutions	1 (Chemical Measurements) 2 (Tools of the Trade)	N/A	5 (Stoichiometric Calculations: The workhorse of the Analyst)
Classical Analytical Methods - A design Perspective on Volumetric Measurement	1 (Chemical Measurements)	N/A	8 (Acid-Base Titrations)
Sample Preparation	27 (Sample Preparation)	N/A	18 (Sample Preparation: Solvent and Solid-phase Extraction)
Instrumental Calibration	4 (Statistics) 5 (Quality Assurance and Calibration Methods)	1 (Introduction)	3 (Statistics and Data Handling in Analytical Chemistry)
Quality Assurance Measures – How Good is the Data?	5 (Quality Assurance and Calibration Methods)	1 (Introduction)	4 (Good Laboratory Practice: Quality Assurance and Methods of Validation)
Instrumental Calibration: Method of Standard Addition	5 (Quality Assurance and Calibration Methods)	1 (Introduction)	17 (Atomic Spectrometric Methods)
Internal Standards	5 (Quality Assurance and Calibration Methods)	1 (Introduction)	17 (Atomic Spectrometric Methods)
Interlaboratory Comparisons	5 (Quality Assurance and Calibration Methods)		4 (Good Laboratory Practice: Quality Assurance and Methods of Validation)
Sampling Error and Sampling Design	27 (Sample Preparation)	Appendix 1	26 (Environmental Sampling and Analysis)

Statistics

Errors in Measurements and their Effect of Data Sets	3 (Experimental Error) 2 (Tools of the Trade)	1 (Introduction)	3 (Statistics and Data Handling in Analytical Chemistry)
The Gaussian Distribution	4 (Statistics)	Appendix 1	3 (Statistics and Data Handling in Analytical Chemistry)
Statistical Tests of Data: The t Test	4 (Statistics)	Appendix 1	3 (Statistics and Data Handling in Analytical Chemistry)
Statistical Tests of Data: The F Test	4 (Statistics)	Appendix 1	3 (Statistics and Data Handling in Analytical Chemistry)
Linear Regression for Calibration of Instruments	4 (Statistics)	1 (Introduction)	3 (Statistics and Data Handling in Analytical Chemistry)

Equilibrium

The Importance of Ionic Strength	7 (Activity and the Systematic Treatment of Equilibrium)	N/A	6 (General Concepts of Chemical Equilibrium)
The Importance of Ionic Strength – Biochemistry Focus			
Activity and Activity Coefficients	7 (Activity and the Systematic Treatment of Equilibrium)	N/A	6 (General Concepts of Chemical Equilibrium)
Multiple Equilibria	6 (Chemical Equilibrium) 12 (Advanced Topics in Equilibrium)	N/A	11 (Precipitation Reactions and Titrations)
pH of Solutions of Strong Acids and Bases	8 (Monoprotic Acid-Base Equilibria)	N/A	7 (Acid-Base Equilibria)
Acid-Base Distribution Plots	8 (Monoprotic Acid-Base Equilibria) 9 (Polyprotic Acid-Base Equilibria)	N/A	7 (Acid-Base Equilibria)
Acid-Base Distribution Plots-Biochemistry		N/A	
Acid-Base Distribution Functions	9 (Polyprotic Acid-Base Equilibria)	N/A	7 (Acid-Base Equilibria)
The Buffer Zone: What is a Buffer and in what pH range is it effective?	8 (Monoprotic Acid-Base Equilibria)	N/A	7 (Acid-Base Equilibria)
The Buffer Zone – Biochemistry Focus			

When Acids and Bases React: Laboratory	1 (Chemical Measurements) 10 (Acid-Base Titrations)	N/A	8 (Acid-Base Titrations)

Electrochemistry

Electrochemistry: The Microscopic View of Electrochemistry	13 (Fundamentals of Electrochemistry)	22 (An Introduction to Electroanalytical Chemistry)	12 (Electrochemical Cells and Electrode Potentials)
Electrochemistry: Calculating Cell Potentials	13 (Fundamentals of Electrochemistry)	22 (An Introduction to Electroanalytical Chemistry)	12 (Electrochemical Cells and Electrode Potentials)

Spectrometry

The Beer-Lambert Law	17 (Fundamentals of Spectrophotometry)	6 (Introduction to Spectrometric Methods) 13 (An Introduction to Ultraviolet-Visible Molecular Absorption Spectrometry)	16 (Spectrochemical Methods)
The Beer-Lambert Law – Biochemistry Focus			
Atomic and Molecular Absorption Processes	17 (Fundamentals of Spectrophotometry)	6 (Introduction to Spectrometric Methods)	16 (Spectrochemical Methods)
Introduction to Spectrometers	17 (Fundamentals of Spectrophotometry) 19 (Spectrophotometers)	7 (Components of Optical Instruments)	16 (Spectrochemical Methods)
Infrared Spectrocopy: Vibrational Modes	19 (Spectrophotometers)	16 (An Introduction to Infrared Spectrometry)	16 (Spectrochemical Methods)
Mass Spectrometry: Interpretation	21 (Mass Spectrometry)	20 (Molecular Mass Spectrometry)	22 (Mass Spectrometry)

Chromatography and Separations

Introduction to Chromatography	22 (Introduction to Analytical Separations)	26 (Introduction to Chromatographic Separations)	19 (Chromatography: Principles and Theory)
Introduction to Gel Filtration – Biochemistry Focus			

Band Broadening Effects in Chromatography	22 (Introduction to Analytical Separations)	26 (Introduction to Chromatographic Separations)	19 (Chromatography: Principles and Theory)
Gas Chromatography or HPLC? Which do you Choose?	22 (Introduction to Analytical Separations) 23 (Gas Chromatography) 24 (High-Performance Liquid Chromatography)	27 (Gas Chromatography) 28 (Liquid Chromatography)	20 (Gas Chromatography) 21 (Liquid Chromatography and Electrophoresis)

D.C. Harris, *Quantitative Chemical Analysis,* 8th Edition, 2009, W.H. Freeman: USA.

Skoog, Douglas A., F. James Holler, Stanley R. Crouch, *Principles of Instrumental Analysis,* 6th Edition, 2007 Cengage: USA.

Christian, Gary D. *Analytical Chemistry,* 9th Edition, 2013, John Wiley and Sons: USA.

Prepared by Elizabeth Jensen, and Caryl Fish

Printed in the USA
K044011SCI011817   01S29053000000001771